Let's Go Buggy!

Troy Corley

Let's Go Buggy!

The Ultimate Family Guide to Insect Zoos and Butterfly Houses

2002
Thanks for "going buggy" with me
Bugfully yours,
Troy

Corley Publications
P.O. Box 16969
Encino, CA 91416-6969
www.letsgobuggy.com

Copyright 2002 by Troy Corley. Printed and bound in the United States of America. All rights reserved. No part of this book may be reproduced or transmitted in any form or by any means, electronic or mechanical, including photocopying, recording, or by an information storage and retrieval system—except by a reviewer who may quote brief passages in a review to be printed in a magazine, newspaper or on the Web—without permission in writing from the Publisher. For information, please contact Corley Publications, P.O. Box 16969, Encino, CA 91416-6969.

Although the author and publisher have done thorough research to ensure the accuracy and completeness of information contained in this book, we assume no responsibility for errors, inaccuracies, omissions, or any inconsistency herein. Any slights of people, places or organizations are unintentional. First edition 2002.

ISBN: 0-9706242-0-4

LCCN: 2002090942

Attention corporations, professional organizations, universities and colleges: Quantity discounts available on bulk purchases of this book for educational, gift purposes, or as premiums. For information, please contact Corley Publications, P.O. Box 16969, Encino, CA 91416-6969; sales@letsgobuggy.com.

Book Design: Mike Diehl – www.MikeDiehl.com
Cover Photo: Dylan Altman of Saugus, California. Used by permission. Licensed by the Natural History Museum of Los Angeles County. Photographed by Dick Meier, staff photographer for the Natural History Museum of Los Angeles County.
Back Cover Photo Credits: Photo on the left is the Monsanto Insectarium's Mary Ann Lee Butterfly Wing at the St. Louis Zoo. Copyright St. Louis Zoo. Photographed by Chuck Dresner. Photo on the right is of an African Giant Millipede at the Minnesota Zoo. Copyright Minnesota Zoo. Both used by permission.

**To
Emelia and Tristan
my two bugatists**

and

**To Ron
a real human being**

Acknowledgements

Throughout the metamorphosis of this book, there have been certain individuals whose interest, enthusiasm and guidance have helped the book to emerge. I want to thank Gina Ward, former Communications Director, Natural History Museum of Los Angeles County; James R. Gilson, Vice President, Natural History Museum of Los Angeles County; Mary Baerg, Natural History Museum of Los Angeles County; Dr. Art Evans, Research Associate, Smithsonian Institution; Dick Meier, photographer at the Natural History Museum of Los Angeles County; Kecia Altman, mother of Dylan, whose photo graces the cover; Ellen Griffee and Karen Lubeniecki, former public relations associates for the Association of Science-Technology Centers; Heather Cooper, public relations for the Smithsonian Institution's National Museum of Natural History; Nate Irwin, Director, Otto Orkin Insect Zoo; Brent Karner, Director, Ralph M. Parsons Insect Zoo at the Natural History Museum of Los Angeles County; Chad Lynde & Ken McFall, San Bernardino County Museum; Dennis Schatz, Pacific Science Center; Jodi Butler, *FamilyFun* magazine; Roberta Sandler, author and friend; Susan Peterson, author and independent publisher; and Mike Diehl, commercial artist. Many thanks to the dozens of entomologists, museum educators, zoo directors and public relations representatives I have talked to in order to glean the best information about these buggy habitats. Thanks to all the friends and family who cheered me on.

I am especially grateful to my two wonderful children, Tristan and Emelia Cassel, who have gone buggy with me for many years, and to Ron O'Brien, whose insight and devotion made this book possible.

Special thanks to my parents, Pat and Iris Corley, whose love and support let me go buggy in my youth and more recently, with this book. And Mom, I promise not to bring any more grasshoppers in the house!

Table of Contents

Introduction 8

A Guide to Using This Book 10

Go Buggy
A guide to insect zoos and butterfly houses 13

Bugged Out
A swarm of bug festivals and other infested events 183

Bug Bytes
A Web of bug cams and Internet insect sites 223

Be a Bugatist
A host of buggy resources 235

Bug Bites
A guide to state insects and pet bugs 243

Bug Buzzwords
A glossary of bug biology 247

Index 249

Introduction

My Chock Full O' Nuts™ coffee can was chock full of giant green grasshoppers, caught on a warm summer day outside our garden apartment complex in Bayside, Queens, New York. My mother hated grasshoppers and would not let me bring them inside until a neighbor convinced her that she should. The bugs bounced in that grass filled tin until I forgot to snap the lid back on.

My mother's screams were so loud that the whole neighborhood heard her. The hoppers had been sprung. They clung to the walls. They jumped on my mother. They hopped onto my younger brothers and sisters who started screaming too. Later my mother would say, "I like ladybugs, but I can't stand grasshoppers. You can make pets out of ladybugs, but not grasshoppers."

Thirty years later, I remembered the grasshopper caper as I sat in my car with 1,500 ladybugs crawling everywhere. My daughter, Emelia, eager to see so many bugs at one time had grabbed the tub just as I had peeled back the lid to give her a peek. The ladies quickly poured out. At first I was angry. More than a thousand bugs had spilled in my lap and in my car. But only a bug-loving mom could laugh as several neighborhood kids climbed in the minivan, scooped up the beetles and deposited them into plastic zippered bags.

Our family has been going buggy for years now. Emelia and her older brother Tristan have investigated Southern California's backyard bugs including Monarchs, Pill Bugs and the indomitable Argentine Ants. We've explored entomology at the **Natural History Museum of Los Angeles County's Insect Zoo** and joined the buzzing crowds at the museum's annual insect fair.

As a journalist and research fanatic, I decided to track down other places with live insects and other invertebrates. To my surprise, there were several in the U.S., even live butterfly houses where you could be surrounded by a bevy of beauties. (This was before our own museum opened a seasonal butterfly pavilion every year). My research became an article for *FamilyFun* magazine's December 1996 issue and I knew then that the topic had the makings of a book.

As my family continued to go buggy with Malaysian Millipedes, Painted Lady Butterflies and Honey Bees, I entered the daunting world of book publishing. After trying to create a buzz about *Let's Go Buggy!* with traditional publishers, I decided to metamorphose from an independent writer into an independent publisher. Corley Publications has emerged from its chrysalis and *Let's Go Buggy!* is the first of many books. (Look for the forthcoming *Bug Bytes: The Ultimate Guide to Insects and Other Bugs on the Web*).

Bugfully yours,
Troy Kathleen Corley

A Guide to Using This Book

While "going buggy" with this guidebook, please note that I use the term "bug" loosely. Kids point to all types of invertebrates, from six-legged insects to eight-legged arachnids to sticky snails and declare them to be BUGS. *Let's Go Buggy!* includes Bug Buzzwords (a glossary) to help dissect bug biology and explain the differences between a "true bug" and a butterfly, an arachnid and an arthropod, and a chrysalis and a cocoon.

For the purposes of this book, I have shed some traditional grammatical rules. Every bug and butterfly name is capitalized, including the ordinary invertebrate term such as cockroach, ant or butterfly when it's part of a complete name. For example, Madagascar Hissing Cockroach, Velvet Ant, Zebra Longwing Butterfly. When referring to groups of invertebrates in general terms, the type of invertebrate is in lowercase. For example, cockroaches, ants, butterflies. I have also capitalized the names of plants used in butterfly houses for easier identification.

The main chapter of the book profiles more than 100 places where you can go buggy in the U.S. There are insect zoos at natural history museums such as my favorite museum in Los Angeles, insectariums at traditional zoos like Missouri's **St. Louis Zoo**, observation Honey Bee hives including the one at the **Creative Discovery Museum** in Chattanooga, Tennessee, and specialized collections such as the **Arachnid Exhibit** at Kentucky's **Louisville Zoo**.

For many families, the live butterfly exhibits are irresistible. There are magnificent glass-domed pavilions like **Callaway Gardens** in Georgia, stunning seasonal enclosures and bountiful butterfly gardens. Wear brightly colored clothing, a hint of perfume or fragrant lotion and a gorgeous butterfly may land on your shoulder or your nose!

It's important to note for all of the bug and butterfly places that times, dates, and admission prices are subject to change. My advice: call before you go so you're not unpleasantly surprised. Change is not only possible but also probable when you have live bugs and butterflies on exhibit. Species come and species go. Colonies crash, government regulations are modified and institutions metamorphose. What's fun about change is that new species you haven't seen may be on display the next time you visit. For example, many museums scoop up local insects to put on display in the summer—you'll be amazed at what could be crawling around your own backyard.

I have not attempted include every butterfly *garden* on U.S. soil. Butterfly plots are popular with botanical gardens and there are few that do not devote a portion of their blooms to attracting these winged wonders. What I have done is include some larger botanical gardens that publicize their butterfly areas and to include butterfly gardens that are a part of institutions such as science centers where gardening is not their primary business.

There are undoubtedly live insect collections and butterfly houses that I have missed. New ones are blossoming every year and butterfly enclosures, in particular, are springing up from coast to coast. Not all are created equal. Some hand-raise local species, some import tropical varieties, some have both, while others let in what ever happens to be around. It's especially fun when a flutterby place has a visible emerging area so visitors can watch as butterflies and moths come out from their chrysalides and cocoons.

For extended buggy fun, turn to the Bugged Out festivals chapter. If you take the kids on a road trip, you may find some unusual goings on like the **Tarantula Festival** in the historic town of Coarsegold, California, or the magnificent **Texas Butterfly Festival** in Mission or the **Woollyworm Festival** in Banner Elk, North Carolina. Again, call ahead to check if the festival is still planned and to verify dates, times and costs.

The back of the book features resources for parents, educators and kids who want to go buggy beyond visiting zoos or festivals. You'll discover organizations to join, more books to read, where to watch bug cams on the Web and some simple ideas on how to keep pet bugs at home or in the classroom.

If you've got some buggy ideas to share, good Web sites to view, new butterfly houses to visit, give me a buzz at bugmom@letsgobuggy.com. Be sure to click on www.letsgobuggy.com for updates on insect zoo openings, featured festivals and pet bug care sheets.

Enjoy the book and Happy Bugging!

—**The Bug Mom**

CHAPTER ONE

Go Buggy

**A guide to insect zoos
and butterfly houses**

● ● ● ● ● ●

Kids love bugs. They are fascinated by these tiny creatures that creep, crawl, fly and buzz. Give kids a chance to touch live bugs or watch butterflies up close and watch how fast they clean up their room.

So, where do you go to see live bugs other than your backyard? There are dozens of places in the U.S. squirming with native and exotic bugs and butterflies, from small buggy exhibits at science centers to crystal butterfly palaces. This chapter gives the details on 140 of them. That's a lot of bugs!

Alabama

Biophilia Nature Center

12695 County Rd. 95, Elberta, AL 36530
251-987-1200
www.biophilia.net

After sailing the seas on their handcrafted boat, the Daedalus, Californians Fred and Carol Lovell Saas permanently tied down in Robert's Bayou. Since 1989 they have been turning 20 acres into an environmental haven for native plants and wildlife.

They started by ripping out foreign flora and replacing it with native Southeast coastal plain vegetation. More than 7,000 hand-planted trees, 300 native plant species including 90 varieties of wildflowers have transformed this acreage into a nature center. The blossoming bounty of plant life is a boon to local butterflies, which easily find their way to the center and into its open-door butterfly house. The screened greenhouse is rarely heated except during a freeze and never cooled. The couple let nature take its course and during peak butterfly season, you never know what you'll find inside.

Native carnivorous plants are also found here as the Saas couple tries to bring back Pitcher Plants and Venus Flytraps. These natives plus butterfly host and nectar plants as well as hummingbird-friendly plants are available at the **Biophilia Native Nursery**. The center also has a caterpillar nursery, a library and a bookstore.

Chart a course on the couple's 50-foot sailboat through local bays and bayous. All tour profits help fund the nature center.

Hours: Call for a reservation.

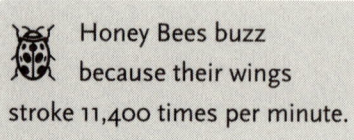
Honey Bees buzz because their wings stroke 11,400 times per minute.

Admission: $5 per person. Biologist-guided tours available for groups. Call for prices on sailboat excursions.

Incredible Invertebrates

Montgomery Zoo

20301 Coliseum Parkway, Montgomery, AL 36110

334-240-4909

http://zoo.ci.montgomery.al.us

Want to go buggy for your birthday? Then book your bug time now for **Incredible Invertebrates**, the zoo's 45-minute educational program. Give kids a close-up look at a scorpion as big as your hand or pet a Giant Madagascar Hissing Cockroach. Other exotic invertebrates include a Black Widow, Orange Knee Tarantula, Carolina Mantid, native stick bugs and a Wolf Spider. This bugfully delightful program is for groups who make a reservation. The program can also go on the road and come to you, within about 100 miles of the zoo. Learn how to identify some of Alabama's winged wonders in **Butterflies of Alabama**, another bugged-out program that includes a slide presentation and some live specimens.

Reserve a time about two weeks in advance. The zoo can prepare special lunch combos with a hot dog, corn dog, hamburger or pepperoni pizza for $3.50 to $4.50 per person with advanced lunch reservations.

Hours: (Zoo) 9 a.m. to 5 p.m. daily.

Admission: (Zoo) $6 adults, $4 ages 3–12, $3 seniors (65+), free for ages 2 and under. (Educational programs) $40 for up to 40 and $50 up to 75 people at the zoo; cost is higher for road trips.

Tessman Butterfly House

Huntsville-Madison County Botanical Garden

4747 Bob Wallace Ave., Huntsville, AL 35805
256-830-4447
www.hsvbg.org

Explore the wonders of vibrant spiraling colors under a shade cloth roof when more than 30 species of North American butterflies take flight at the 122-acre gardens. Zebra Longwings, Viceroys, Monarchs, swallowtails and fritillaries flutter throughout the butterfly-friendly landscape.

The summertime shower of shimmering wings includes a showcase of giant sculpted butterflies, created by local artists. Discover these 4.5' and 8' wonders among the flowers and shrubs in the gardens. Be sure to visit the butterfly garden, **Dogwood Trail** and the **Nature Trail** studded with wildflowers for more butterfly beauty.

Help the **Butterfly House** fly by sponsoring a butterfly to be released in honor or in remembrance of someone special. In addition to releasing the flying flower into the house, the recipient gets a special watercolor print and a card. Cost is $25 per butterfly released.

Hours: (Gardens) 9 a.m. to 6:30 p.m. Mon.–Sat., 1 p.m. to 6:30 p.m. Sun., April to October; 9 a.m. to 5 p.m. Mon.–Sat., 1 p.m. to 5 p.m. Sun., November to March. (Butterfly House) May to September.

Admission: $6 adults, $5 seniors, $3 ages 3–18.

> The Queen Alexandra Butterfly is the world's largest with an 11-inch wingspan.

Alaska

Butterflies of Alaska

University of Alaska Museum

907 Yukon Dr., Fairbanks, AK 99775
907-474-7505
www.uaf.edu/museum/ento

Alaska, alas, is one of those states that does not have a live collection of bugs or butterflies that's accessible to the public. It does however, have a fairly new entomology collection at this university museum that features a **Butterflies of Alaska** display. The museum was recently given a grant from the National Science Foundation as part of its **Arctic Archival Observatory**, so look for its insect collection of about 100,000 aquatic bugs to expand. Luckily, the museum does hold an annual **TOTE** (Totem Ocean Trailer Express) **Family Fun Fest**, which includes several hands-on activities for kids including an opportunity to explore bug life (see Bugged Out chapter).

Hours: 9 a.m. to 5 p.m. weekdays, Noon to 5 p.m. weekends, mid-September to mid-May; 9 a.m. to 7 p.m. mid-May to mid-September. Closed Thanksgiving, Christmas and New Year's Day.

Admission: $5 adults, $3 ages 7–17, $4.50 seniors (60+), free for ages 6 and under.

Arizona

Arizona-Sonoran Desert Museum

2021 N. Kinney Rd., Tucson, AZ 85743
520-883-1380
www.desertmuseum.org

Stark, barren and bare are words you leave behind when the desert's beauty is unveiled at this extraordinary place outside Tucson. Part zoo, part botanical garden, this 100-acre mostly outdoor museum opens up your eyes to the wonders of desert life from rare Gray Wolves to Prairie Dogs to hummingbirds.

Among the 300 species of animals living here are vibrant invertebrates, many of which are coveted by collectors. The insect zoo "regulars" include Black Widows and Arizona Browns, the two most dangerous spiders in Arizona. There are Giant Hairy Scorpions, Tail-Less Whip Scorpions, Giant Desert Centipedes, Vinegaroons and several tarantula species such as the Desert Blonde. The Sonoran Desert native arthropod exhibits rotate, so you might see Praying Mantids, shiny emerald green Leaf Chafers, Dung Beetles, Dynasty Beetles and various grasshoppers. Of particular interest are the Giant Mesquite Bugs. "True bugs" because they're members of the Hemiptera order, meaning half-wing, the red and black insects munch on Mesquite trees.

Even a desert attracts butterflies and more than 50 species have been seen at the museum. In the **Desert Garden** area a butterfly garden entices more than 30 species to visit on a regular basis: Queens, Gulf Fritillaries, Cloudless Sulphurs, Southern Dogfaces, Metal Marks, Tropical Buckeyes (related to Buckeyes only darker) and Dusky Scallop Wings, a Skipper Butterfly whose folded wings reveals a scalloped edge.

Watch solitary Native Bees pollinate relatives of tomato plants and nightshade in a small garden. Non-honey makers, the females build a burrow nest and raise the young themselves.

During the hot summer the museum stays open until 10 p.m. on Saturdays, so you can see the wildlife that blossoms at night, especially in the moth garden. Look for Giant Sphinx Moths around perfume emitting flowers like Evening Primroses that bloom after 5 p.m. Summer nights are perfect for insect hunts, which the museum holds regularly on Saturdays. You can also go on scorpion hunts using black lights on sticks like carpet sweepers.

A colony of Harvester Ants lives in the **Desert Grassland** habitat. These winged insects forage for seeds and can alter the habitat they live in by harvesting too many. Look for them in a see-through exhibit in the side of an embankment. This type of natural setting will be expanded in a new exhibit called **Life on the Rocks**, opening in fall 2002 with live animals and invertebrates.

Try to time a fall visit for September when the museum holds its annual **Butterfly Festival**.

Hours: 8:30 a.m. to 5 p.m. October to February; 7:30 a.m. to 5 p.m. March through September; until 10 p.m. Saturdays, June through September.

Admission: $9.95 adults November through April, $8.95 adults May through October, $1.75 ages 6–12, free for ages under 6.

Sonoran Arthropod Studies Institute

2114 W. Grant Rd., #39, Tucson, AZ 85745
520-883-3945
www.sasionline.org

An exceptional educational center, SASI is well known among entomologists and invertebrate enthusiasts. Located within Pima County's 18,422-acre **Tucson Mountain Park**, the non-profit institute was founded by arthropod aficionado Steve Prchal in 1986.

SASI invites the public to visit its facilities and dozens of invertebrate species on **Community Days**, held the fourth Saturday every month except December. SASI's classroom is transformed into a temporary exhibit on these days, showcasing live arthropods from the teaching collection and pinned specimens from the research collection.

See several species of tarantulas, Black Widows, Brown Spiders, Giant Crab Spiders, Desert and Giant Desert Centipedes, Bark Scorpions, Giant Hairy Scorpions, Desert and African Millipedes, Vinegaroons and Velvet Mites. The center maintains more than two dozen species of ants such as Honey, Leaf Cutters and Acrobats and at times ant colonies are for sale. Insects include Darkling Beetles, Antlions, Giant Water Bugs, Diving Beetles, Caterpillars, Longhorn Beetles, Velvet Ants, Hissing Cockroaches, dragonfly larvae, grasshoppers and Scarab Beetles. There are also crustaceans such as isopods, crayfish and Hermit Crabs.

> Arachnophobia, fear of spiders, is the number one phobia in the world.

These arthropods are available to elementary school classrooms for outreach programs. Become a member of SASI and you can visit all these critters Tue.–Sat. by appointment.

Be sure to take a walk on the trails and visit the small butterfly garden that was once an ugly parking lot. The garden attracts species such as Queens, Sleepy Oranges, Cloudless Sulphurs, Mexican Sulphurs and Acacia Skippers.

If you go totally buggy over arthropods, consider attending SASI's annual **Invertebrates in Captivity Conference**. Learn about caring for invertebrates, educational programs using invertebrates, discover more about insect zoos and butterfly houses and conservation issues. Besides presentations to listen to, there are several insect field trips scheduled.

Hours: (Community Days) 9 a.m. to 3 p.m. fourth Saturday of each month except December. (Members) 9 a.m. to 4 p.m. Tue.–Sat. by appointment.

Admission: (Community Days) Free. (Annual membership) $25 individual, $40 family, $100 contributing or professional, $500 life.

Bug House

Phoenix Zoo

455 North Galvin Parkway, Phoenix, AZ 85008
602-273-1341
www.phoenixzoo.org

On the zoo's **Discovery Trail**, which features **Harmony Farm** and petting zoo, you'll find a small **Bug House**. Essentially a cubicle with terrariums embedded in the wall, the **Bug House** has about 10 different species of invertebrates: Scarab Beetles, Black Widows, scorpions, millipedes, Madagascar Hissing Cockroaches, Bird Eating Tarantulas, Giant Desert and Sonoran Centipedes.

The zoo sponsors a variety of buggy workshops where you can watch the Blue Death Feigning Beetle play dead, touch a Giant African Millipede and view a variety of scorpions from the Hairy to the Barbed to the Tail-Less Whip variety.

If you remember the brave ants from the movie *A Bug's Life*, then you'll want to see the zoo's Leaf Cutter Ant colony on the **Tropics Trail**. Located in an exhibit resembling a Chilean rain forest, the ants' nest can be viewed through several glass compartments embedded in the wall. Watch as ants haul leaves and flowers on their backs, look at the mound where they stash their fungus and observe the workers and the warriors. There's a larger Leaf Cutter Ant colony in the zoo auditorium.

Hours: 9 a.m. to 5 p.m., September to May; 7 a.m. to 12 p.m. and 6 p.m. to 9 p.m. June 1 to July 31; 6 p.m. to 9 p.m. in August only.

Admission: $10 adults, $9 seniors, $5 children ages 3–12, free for ages 2 and under, September to May; $6 adults, $5 seniors, $3 ages 3–12 June to July; $3 all ages August only.

The Tarantula Grotto

Heritage Park Zoo

1403 Heritage Park Rd., Prescott, AZ 86301
928-778-4242
www.heritageparkzoo.org

Most of the native animal species at this 10-acre zoo were injured or orphaned. The exotic endangered species were captive born surplus from other zoos or animal facilities. In the midst of the mammals like tigers, jaguars and pumas, the zoo has a unique residence for tarantulas of the world, located across from the mountain lions.

Fondly called the **Tarantula Grotto**, the exhibit looks like an adobe hut. Inside, 17 species of the Family Theraposidae are on display, *in the dark*. The spiders are in terraria in as natural a setting as possible, so no bright lights. As you walk in, look for the flashlights so you can peek at these eight-legged crawlers. The tarantulas range from the biggest in the world—the Goliath Bird Eater, to the three-inch White Collared Tarantula.

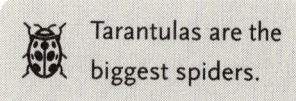

Tarantulas are the biggest spiders.

Among the species: Togo Starburst from Africa, Cobalt Blue from Asia, Greenbottle Blue from Venezuela, Desert Blond from Southwestern U.S. and the King Baboon from Kenya. Look for the big horn on the Straighthorned from Zimbabwe.

Hours: 9 a.m. to 5 p.m. May 1 to October 1, 10 a.m. to 4 p.m. November 1 to April 30.

Admission: $5 adults, $4 seniors, $2 ages 3–12, free ages under 3.

Katydid Insect Museum

5060 W. Bethany Home Rd. #4, Glendale, AZ 85301
623-931-8718
www.insectmuseum.com

There's a lot of buzzing going on in this Southwestern desert town, and it's mostly from a bug loving grandma and her museum. Nedra Solomon has put her bug collection of dead and live arthropods on display to help dispel entomophobia among the public at large. Grandma Nedra came to appreciate bugs while running her Heritage Pest Control exterminating business for almost 30 years.

Fascinated by bugs because of her business, Nedra took to taking local insect safaris with her husband Al and started a nice collection. That is until a two-year-old grandchild smashed the little suckers with a fly swatter as they sat on Nedra's dining room table. The collection has been reconstructed since then—there are more than 4,700 specimens on display. Most of the museum's bugs are native to Arizona but there are live exotics such as Madagascar Hissing Cockroaches. Luckily the grandkids are old enough to know not to swat them!

In fact, Nedra's grandchildren are a big part of the museum. Sections of this collector's dream are based on themes using the names of each of her grandchildren and the museum itself gets its name from oldest grandchild Katherine "Kate" Smith. For birthdays, Mother's Day and other holidays, they present Nedra with buggy presents such as Giant Walking Sticks, Goliath Beetles and Atlas Moths, which are then added to the museum.

You never know what you'll see live here. Perhaps an Emperor Scorpion or a Tarantula Hawk. There are millipedes, mealworms, a Wolf Spider, a Black Hole Spider that designs a strange looking web according to Nedra, a variety of tarantulas, Water Scorpions, Desert Stink Beetles, Diving Water Beetles and Whirligigs. You'll also find live turtles, Fire Belly Toads and a Bearded Dragon.

If you've got a specimen to donate, Nedra is happy to accept.

Hours: 11 a.m. to 4:30 p.m. Mon.–Fri.

Admission: $3 adults, $2 students and seniors, $1 ages 5–11, free for ages 4 and under.

Arkansas

Bug Zoo

Museum of Discovery

500 President Clinton Ave., Suite 150, Little Rock, AR 72201
800-880-6475
www.amod.org

Colorful surroundings, well-lit terrariums and kid-friendly access make this children's museum's **Bug Zoo** a success. The green and yellow banner waving above the glass bug houses, bright blue walls and sunny yellow chairs, are welcoming signs in this **World of Forests** exhibit. The terrariums filled with invertebrates are set on two low shelves with seats stationed in front of them so you can sit and watch these representatives of the world's diversity for a while.

Brazilian Cockroaches climb on a tree limb busily eating their fruits and veggies in a rain forest environment. There are tarantulas, centipedes, millipedes, Australian Walking Sticks, Emperor Scorpions, Water Bugs, Cactus Beetles and Velvet Ants. Look through a microscope at some miniscule bugs such as protozoa. During the warmer months, the museum brings in a variety of native bugs such as katydids, local walking sticks and beetles.

Downstairs, just past the zoo, you'll find an observation beehive busily working in an exhibit that resembles a tree stump. Watch as the bees go in and out of a tube through an exit in the wall.

Sponsored by the Terminex pest control company, the **Bug Zoo** also hosts an annual **Bug Fest** every July.

Hours: 9 a.m.–5 p.m. Mon.–Sat. Closed Sun. except for special events.

Admission: $5.95 adults, $5 seniors (65+), $5.50 ages 3–12, free for ages under 3.

Arthropod Museum

University of Arkansas

Agricultural Building, Fayetteville, AR 72701

479-575-2482

www.uark.edu/depts/entomolo/museum/overview.html

It's a small, small world but this museum has the largest collection of bugs in the state. Located on the third floor of the **Agricultural Building**, the museum is the only place in Arkansas where you can see an expansive collection of bugs.

Comprised of primarily preserved specimens, the museum holds 750,000 individual bugs representing 25,000 species. It's one of the two best collections of insects native to the Ozark and Ouachita Mountains. This is the place to go when you find a bug and want to know what it is and what it does. The museum is small so call first to make sure the museum can accommodate you.

Despite being a center of dead bugs, the museum sponsors a lively insect festival every October with live bugs, an observation Honey Bee hive and cockroach races (see Bugged Out chapter).

Admission: Free.

California

Ralph M. Parsons Insect Zoo

Natural History Museum of Los Angeles County

Exposition Park, 900 Exposition Blvd., Los Angeles, CA 90007

213-744-DINO

www.nhm.org

Follow the marching ants on the museum walls to the **Children's Discovery Center** where the insect zoo occupies the second floor. Jungle Nymph Sticks, Green African Flower Mantids and Elephant Beetles are among the 70 species of insects residing in one of the largest live insect displays in the U.S. New **Insect Zoo** director Brent Karner has revitalized the zoo's live collection, doubling the number of species in the last two years. The zoo is dynamic—you never know what new bugs have been hatched in the zoo's rearing facilities until you pay a visit.

What's the Buzz?, an interactive exhibit, demonstrates the sound-producing mechanisms of crickets, cicadas and katydids and the flashlight effects of fireflies. Open a life-sized double-door refrigerator to reveal what bugs pack for lunch. Use the Bioscanner to zoom in on the insects crawling and swimming in a combination terrarium/aquarium. The scanner lets you magnify the bugs up to 100 times; you can also see the image on a large video monitor. This changing exhibit has included Backswimmer Bugs, Carpenter Ants, Dampwood Termites and Praying Mantids. When you visit who knows whose hairy legs you'll see up close and personal.

On the first floor observe the wiggly wonders of an earthworm farm; touch mealworms, beetles and sow bugs; see butterfly chrysalides and examine dozens of kid-size bug display cases. Butterflies fly in for the spring and summer months when the museum opens it annual **Pavilion of Wings** live butterfly habitat featuring hundreds of tropical and native species (see page 27). In May the crowds come crawling in for a weekend full of bugged-out fun at the largest insect fest in the West. About 10,000 infest the museum bringing live bugs and specimens for sale, cooked bugs to eat and a bevy of buggy crafts and activities.

Hours: 9:30 a.m. to 5 p.m. Mon.–Fri., 10 a.m. to 5 p.m. Sat.–Sun. Closed Independence Day, Thanksgiving, Christmas and New Year's Day.

Admission: $8 adults, $5.50 seniors and students with I.D., $2 ages 5–12, free for ages under 5. Includes Natural History Museum, Insect Zoo and Children's Discovery Center. Free on first Tuesday of the month.

Pavilion of Wings

Natural History Museum of Los Angeles County

Exposition Park, 900 Exposition Blvd., Los Angeles, CA 90007

213-763-DINO

www.nhm.org

Every April, the butterflies bloom in Los Angeles's **Exposition Park**. The park is home to the museum, which constructs the seasonal butterfly enclosure on the South Lawn. Families line up for a chance to enter the mini paradise and be surrounded by a swirl of color. Populated with more than 20 North American butterfly and moth species, the pavilion has undergone some dramatic changes under the helm of the museum's new **Insect Zoo** director, Brent Karner. The 36' by 90' greenhouse is liberally planted with butterfly-attracting plants. A simple walkway provides easy access so you can loop through the pavilion. Signage lets even the youngest butterfly admirers identify the winged beauties inside: Red Passion Flowers, Question Marks, Giant Swallowtails, Monarchs, Zebra Longwings, American Painted Ladies, Red Admirals. Try to spot the spectacular Luna Moth with its long wingspan and green hue.

As you reach the end of the loop you'll see an emerging box, where chrysalides sit in the shade ready to let go of their flying treasures. At the tip of the inside planting area notice the plants that are not as lush as the rest of the landscape. This weedy looking foliage is food for the caterpillars. The pavilion is one of the few seasonal enclosures that's allowed to keep caterpillars inside the same space where the butterflies fly.

Volunteers and staff members are on hand to answer questions and help you identify species. Before entering and exiting, you'll get a chance to browse through the **Pavilion of Wings Gift Shop** buzzing with butterfly kits, toys, jewelry and books—even this one!

Hours: 9:30 a.m. to 5 p.m. Mon.–Fri., 10 a.m. to 5 p.m. Sat.–Sun., early May to early September. Last entry: 4:30 p.m.

Admission: (Pavilion) $3 adults, $1 children, $2 students/seniors.

Tickets are sold for entry times on the half-hour. (Museum) $8 adults, $5.50 seniors (62+) and students with I.D., $2 ages 5–12, free for ages 4 and under.

World of Life

California Science Center

Exposition Park, 700 State Dr., Los Angeles, CA 90037
323-SCIENCE
www.casciencectr.org

If you still need to go buggy after visiting the **Natural History Museum of Los Angeles County** then walk over to this free state science center next door. There are three species of invertebrates mixed among the technologically outstanding exhibits. You'll find a colony of California Hardwood Termites in the **World of Life** area. The bugs are busy chomping away at a circular slice of an Oak Tree. The display represents symbiosis. Nearby are hundreds of American Cockroaches in a display that explains how different creatures breathe. The museum's **Discovery Room** features a rotating exhibit that demonstrates insect life. At the time of this writing a colony of Madagascar Hissing Cockroaches had that distinction.

Hours: 10 a.m. to 5 p.m. daily. Closed Thanksgiving, Christmas and New Year's Day.

Admission: Free, but donations are requested.

Parking: $6.

Winnick Family Children's Zoo

Los Angeles Zoo

5333 Zoo Dr., Los Angeles, CA 90027
323-644-4273
www.lazoo.org

Head for the **Desert Trail** in the recently renovated children's area of this 133-acre zoo and you'll find some invertebrates alive and kicking—or burrowing inside a cave-like exhibit. The creepy crawlies reside in glass terrariums embedded in the walls. They're a little hard to see due to low lighting, but look closely for tarantulas, scorpions, millipedes and walking sticks. There are also Mexican Beaded Lizards, chuckwallas and a Desert Tortoise.

Around the corner at the hands-on **Riordan Kids' Korner**, you'll probably find one invertebrate species or another on display on a table such as Madagascar Hissing Cockroaches or Vietnamese Walking Sticks. An on-site docent will answer questions and you may even get to touch the critters.

Visit the adjacent **Muriel's Ranch** where you pet freely roaming goats and alpacas and the outdoor **Adventure Theater** where animal keepers present programs with live animals.

Hours: 10 a.m. to 5 p.m. daily. Closed Christmas.

Admission: $8.25 adults, $5.25 seniors (65+), $3.25 ages 2–12, free for ages under 2.

Exploration Station

San Bernardino County Museum

2024 Orange Tree Lane, Redlands, CA 92374
888-BIRD-EGG
www.sbcountymuseum.org

Located in what's known as the Inland Empire, this is the largest museum in the region and the main outpost for learning about nature and science. A host of live animals can be found in a separate building behind the main museum, including a decent display of invertebrates.

More than a dozen species are featured in an unusual exhibit space: 14 small cages with glass frame fronts are embedded in a wooden cabinet about 7' feet wide by 8' high. Despite the height, there are three different steps so kids can see all the bugs. Resident beetles include local Darklings and Iron Clads, Long Horned Cactus Beetles, juvenile Cave Cockroaches and Blue Death Feigning Beetles also known as Arizona Sand Dunes. The last of these beetles are so good at pretending to be dead, that bug curator Chad Lynde has to remind staff members not to throw them out.

Joining these insects are several arachnids: Black Widows caught at the museum; a host of tarantulas including a Mexican Blonde, a local Desert variety, a Rose Hair and a Red Knee; Vinegaroons; and Emperor and California Desert Hairy Scorpions.

The museum also has two Velvet Ant colonies. One, the red and black tufted species, is known as "cow killers" because of their painful sting. The white "Thistledown" species resemble little hedgehogs. These ants aren't true ants—the females (males are usually hidden) only resemble picnic guests. They're really wingless wasps.

Rounding out the live collection are four freestanding aquariums filled with Vietnamese and Australian Leaf Sticks raised at the museum as well as Madagascar Hissing Cockroaches and South American Giant Cave Roaches. The latter cucarachas look like Death's Head Roaches without the distinctive markings and will spritz you with a stinky pepper-like spray.

Interpreters are on hand every day to talk to visitors and often take out bugs for you to see up close. The museum also hosts several bug education programs throughout the year.

Hours: 9 a.m. to 5 p.m. Tue.–Sun. Open these Monday holidays: Martin Luther King, Jr. Day, Presidents' Day, Memorial Day, Labor Day, Columbus Day. Closed Thanksgiving, Christmas, New Year's Day.

Admission: $4 adults, $3 seniors (60+) and students with I.D., $2 ages 5–12, free for ages under 5.

The Hidden Jungle

San Diego Wild Animal Park

15500 San Pasqual Valley Rd., Escondido, CA 92027
760-747-8702
www.sandiegozoo.org/wap/

At the entrance to the vast open spaces of the 1,800-acre park is a lush, tropical greenhouse and aviary aptly named **The Hidden Jungle**. Butterflies and birds fly free within the walk-through enclosure. During the spring and summer hundreds of delicate Pan Tropical wings flutter among the neo-tropical vines and Strangler Fig Trees. This number increases during the annual two-week event in April called **Butterflies & Orchids** when thousands of flying flowers inhabit the man-made paradise. About 200 to 500 newly hatched butterflies, including metallic Blue Morphos, Red Cattle Hearts and yellow-and-black striped Zebras, are released each day during the colorful spring show. The pupae come from rain forest butterfly farms—the park's way of promoting sustainability in those regions.

Famous for its botanical gardens, you'll plenty of plant life inside the park's enclosure. Nectar plants include Pentas, Lantana and some Costa Rican plants. The jungle also supplies Psiguria, a neo-tropical vine that provides both pollen and nectar for Heliconius Butterflies, increasing their life span. During the annual spring event hundreds of white, pink, purple and yellow orchids spill out from under a covered walkway next to the **Animal Care Center** in **Nairobi Village**.

Owls, Mormons and Julias are not the only creatures living in the jungle. To enter, you must walk though a cave-like opening called **The Crevace**. Along the walls are collections of live insects, including a new colony of Leaf Cutter Ants. The colony lives in **The Crevace** but forages along a vine that enters the greenhouse, so you can watch the swiftly moving line of ants carry leaves back to the nest. Look for three stick insect species: Thorny Sticks, Giant Wingless and Australian Spiny Sticks. There are several tarantula species including a Goliath Bird Eater with an abdomen the size of a fist, a Green Bottle Blue Tarantula and a Brazilian White Knee. Depending on the season and how lucky the breeding populations are you may also find Emperor Scorpions; Dead Leaf and Orchid Mantids; leaf insects; Giant African Millipedes; Atlas, Hercules and Stag Beetles; and centipedes.

The jungle is also home to reptiles like the Emerald Tree Boa and the Sinaloan Milk Snake. Twenty bird species such as hummingbirds, Bare-necked Fruit Crows, White-Bearded Manikins and tanagers, fly as freely as the butterflies. New is a Green Wing Macaw that sits on a perch.

Hours: 9 a.m. to 4 p.m. daily.

Admission: $23.95 adults (12+), $16.95 ages 3–11, free for ages 2 and under. Hidden Jungle included in park admission.

Parking: $6.

Natural Treasures: Past and Present

San Diego Natural History Museum

1788 El Prado, Balboa Park, San Diego, CA 92112-1390
(619) 232-3821
www.sdnhm.org

A lot has changed since this museum opened its first exhibits in a hotel in 1912. For one, the vibrant **Bug Zoo** is gone, a result of the expansion and renovation of the museum's 1933 building in Balboa Park. Most, but not all of the invertebrates are hidden behind lab doors. The museum's temporary exhibit, **Natural Treasures: Past and Present**, contains three invertebrate species: Velvet Ants, Giant Hairy Scorpions and Ghost Beetles. The last buggy bunch is similar to Iron-Clad Beetles but have a waxy secretion on their exoskeleton, hence the name Ghost. Find them in glass terrariums along side Speckled Rattlesnakes, Pacific Tree Frogs and Western Banded Geckos.

During the next five years, the museum plans to open several new exhibits designed to re-create living habitats. Woven into some of the exhibits will be live plants and animals, including busy bugs. The museum's **Discovery Room**, now in the early stages of construction, will feature lots of touchables for kids.

In the meantime, after visiting with the trio of live invertebrate species, look for a variety of preserved insect specimens in the exhibit. Housed in antique glass cases, you'll see California's state

> The largest insects were prehistoric dragonflies with 30-inch wingspans.

insect, the California Dogface Butterfly and other fauna: Pale Swallowtail Butterfly, Ten-Lined Beetle, Prionus Beetle and the Digger Wasp.

Hours: 9:30 a.m. to 4:30 p.m., daily, except Thanksgiving, Christmas, and New Year's Day.

Admission: $7 adults, $6 seniors (60+), $5 ages 3–17, free for ages 2 and under.

The Monarch Program Butterfly Vivarium

450 Ocean View Ave., Encinitas, CA 92024
760-944-7113
www.monarchprogram.org

A non-profit organization, the **Monarch Program** studies the behavior and biology of Monarchs. Its vivarium, a 1,200-square-foot netted enclosure, features native species. A favorite field trip for local schools, the butterfly house is open to the public on Saturdays during the warm months. The house is landscaped with Eucalyptus and Ficus trees, flowering nectar plants, butterfly host plants, and has rearing areas for caterpillars to metamorphose. Butterfly and plant lectures are often held in the museum/classroom.

Become a member and volunteer to tag and track migrating Monarchs. You can buy butterfly livestock from the program and payments are considered a donation to the organization if used for educational or scientific purposes. Purchase Monarchs, Queens, Mourning Cloaks, Red Admirals, Painted Ladies, Buckeyes, Anise Swallowtails and California Dogfaces. Larvae range from $2 to $5 each. Pupae cost $4 to $7 each. Purchase the host plants, too, to feed your caterpillars: Milkweed, Willow, Stinging Nettle, Passion Vine and Fennel.

Hours: 11 a.m. to 2 p.m. Saturdays, April to October. Extended hours during summer months.

Membership: $30

Butterflies of the Central Coast

Mariposa de Carmel

P.O. Box 618, Carmel Valley, CA 93924
831-659-4631
www.butterflieslive.com

Almost a state secret, **Mariposa de Carmel** is neatly hidden along the California Coast. A haven for butterflies, the property has been privately "farmed" by Dr. Abraham Kryger since 1995. **Butterflies of the Central Coast**, which provides Monarchs and Painted Ladies for weddings, films and other events, opened on the farm in January 2000. The butterfly business is owned by Michael Mintz who first became a butterfly breeder when he missed the California lifestyle while living in a New York City apartment.

Although not a public venue, the butterfly farm welcomes school groups and other visitors to tour the private paradise if an appointment is made. Once there, you'll be guided though the 625-square-foot walk-though vivarium where Mintz hand-raises his stock from the caterpillar stage. One of the farm's goals is to have 10 or more species inside the vivarium so visitors can view nature's variety up close. John Lane, a world famous lepidopterist, spent some time living in a trailer on the property several years ago, and identified more than 20 butterfly species here. Sit on a bench and try to identify some of the species while listening to the sounds of the waterfall.

Discover butterfly gardens billowing with host and nectar plants outside the vivarium. There's also a wildflower garden, a compost demonstration site, mushroom garden and greenhouse with host and nectar plants for sale. The greenhouse overflows with Milkweed, which Monarch caterpillars devour daily. Look for the Oaks, Sycamores and native Willow trees that also provide food for several caterpillar species.

The farm has a picnic area so be sure to bring a lunch and spend the afternoon watching for these local beauties.

Hours: By appointment, mid-April to mid-October.

Admission: Suggested donation of $10 adults, $5 children and seniors; includes tour. All monies go to improving the butterfly habitat.

Butterfly World

Six Flags Marine World

2001 Marine World Parkway, Vallejo, CA 94589
707-643-ORCA
www.sixflags.com

Known for its mega monster roller coasters, Looney Tunes™ entertainment and water thrills, this Northern California theme park is also home to a bounty of animals from dolphins and whales to lorikeets and kookaburras to elephants and camels. Amid the animal experiences is **Butterfly World**, the first major walk-through butterfly habitat in the Western U.S.

The 100' by 50' greenhouse-style glass atrium features 500 free-flying butterflies from around the world. More than 700 different species are imported throughout the year, with weekly shipments of chrysalides, so there's always something new to see. The bouquet of butterflies is made up of four major groups: Brushfooted Butterflies, Whites/Sulphurs, Swallowtails and Giant Silk Moths.

Dancing among the tropical plants in the atrium you may find Dragontails, Peacocks, Jezebels, Spangles, Jays, Mimes, Cattlehearts, Lacewings, Longwings, Batwings, Clippers and Queens. Darting around flowering foliage may be Buckeyes, Pansies, Morphos, Monarchs, Malachites, Postmen, Queens, Crows, Zebras and Tigers. Notice how each species has a special way of flying, feeding, perching and roosting.

Watch adult butterflies emerge from their chrysalides in the pupae exhibit, maybe even a select caterpillar or two. Other flying residents of the butterfly house include hummingbirds and Diamond Doves while Koi and goldfish inhabit the waterfall pool. Look for some turtles in their own pool.

Be sure to find the educational displays outside the atrium. Learn about the butterfly life cycle, how butterfly wings get their color, what plants attract native butterflies and more.

Hours: Operates year-round but schedules vary so call first.

Admission: $34 adults, $25 seniors, $17 children 48" and under, free for ages 3 and under.

Insect Zoo

San Francisco Zoo

1 Zoo Rd., Sloat and 45th Ave., San Francisco, CA 94132
415-753-7080
www.sfzoo.org

Besides the Golden Gate Bridge, San Francisco has the distinction of being home to the first permanent insect zoo established in the Western U.S and the third in the entire country. Opened in 1979, the insect zoo houses more than 6,000 arthropods in its own permanent building located in the **Children's Zoo.**

On a recent trip there, more than 30 species of invertebrates were on display, including the unusual Velvet Ants. The densely hairy females of the colony, with black and red tufts, only appear to be ants. They're really wingless wasps. Their sting is so severe that they've earned the nickname of "cow killers."

The zoo also has one of the few Dampwood Termite colonies on display and its American Cockroaches came courtesy of the 1992 movie, *Joe's Apartment.* Joining these roaches are their cousins, the Madagascar Hissing Cockroaches. Big and scary looking, the roaches aren't dangerous and only hiss when disturbed. Cockroaches are beetles and there are more representatives of their family tree here: Darklings, Whirligigs, White-Eyed Assassins, Harlequins, Milkweeds, Giant Water Beetles and African Ground Beetles. Look for the Dung Beetles that push piles of poop around and Skin Beetles, also known as Dermestids that pick the flesh off bones.

Several spiders also call the zoo home including the deadly duo—Black

Widows and Brown Recluses. Tarantula species include: California, Goliath, Indian Ornamental and Mexican Red Leg. Other bugs in this band of invertebrates: Giant African Millipedes, Sri Lankin Mantids, Angular-Winged Katydids, Desert Mantids, McCleary's Spectre Walking Sticks, Giant Walking Sticks, Thorny Phasmids and Giant African Scorpions.

Look for several interactives including "Bee a Bug" vests, microscope tables and touch screens. During weekends bug keepers often bring out tarantulas or millipedes for show and tell.

Hours: (Main Zoo) 10 a.m. to 5 p.m. daily. (Children's Zoo) 11 a.m. to 4 p.m. daily. Extended weekend and summer hours, 10:30 a.m. to 4:30 p.m,. mid-June through Labor Day. Open 365 days a year.

Admission: $10 adults, $7 seniors (65+) and ages 12–17, $4 ages 3–11, discounts for San Francisco residents.

Butterflies!

Turtle Bay Museum

Visitor's Center, 844 Auditorium Dr., Redding, CA 96001
530-243-8850 or 800-TURTLEBAY
www.turtlebay.org

North of Sacramento in the natural beauty of Shasta County, **Turtle Bay Museum** is a monument to the life of the Sacramento River watershed. A complex of several museum buildings, exhibits and interpretive areas, **Turtle Bay** is an ecologist's delight. Currently undergoing an expansion that will transform it into a 300-acre cultural and environmental exploration park, it includes a 35,000-square-foot museum on art, history and nature opening in June 2002.

Not too far from this new main enclave is **Paul Bunyan's Forest Camp**, a life-sized version of a forest camp with wooden play equipment, a mini-Sacramento River exhibit and, **Butterflies!** a seasonal butterfly house. Landscaped by a master gardener, the 100' by 30' shaded enclosure overflows with nectar bearing foliage making several hundred North American flying flowers quite content.

Brimming with Butterfly Bush, Lantana, Pentas and Sunflowers, the misty haven entices species such as Zebra Longwings, Atalas and Gulf Fritillaries to playfully showoff. Look for Orange Banded and Clouded Sulphur Butterflies dancing around the Monkey Flowers, Yarrow and Stone Cups. You may get to see up to a dozen butterfly species including Painted Ladies, Mourning Cloaks and Giant Swallowtails flitting about the Spider Flowers, hanging baskets and aquatic plants. Also look for majestic Monarchs and the state's official insect, the California Dogface Butterfly.

Try to find the towering Flowering Tobacco, which grows as tall as a person and slightly smells of nicotine—butterflies are highly attracted to it. Also look closely at the young Conifers growing inside the house. If you see something that appears to be a dead branch, examine it carefully to discover a blanket of tangerine-colored Julias Butterflies roosting.

On the average the butterfly house tries to maintain 500 winged wonders, according to Dr. Lee Simons curator of natural science. On a good warm day those numbers could swell to over 1,000. Visitors are encouraged to ask questions and several volunteers and staff members are there daily to help people identify and learn about butterflies.

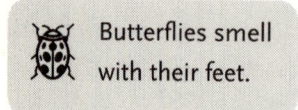

> Butterflies smell with their feet.

See the scientific staff in action at the **Chrysalis Shed** which holds the 400 to 500 pupas purchased from butterfly suppliers in California, Texas and Florida. A few times each day you can watch as butterfly chrysalides are hot-glued onto hanging boards. A glass front on a display box let's you observe a group of chrysalides at all times and early morning you'll see quite a show as butterflies emerge. At times, staff members may also bring out a dissecting scope so visitors can get a magnified view of eggs or caterpillars that have gotten into the house.

Hours: (Butterfly House) 10 a.m. to 5 p.m. late May to late September. (Museum) 10 a.m. to 5 p.m. daily.

Admission: $4 adults, $2 ages 4–17, free for ages under 4.

The Daniel Boone Butterfly Palace
(proposed)

P.O. Box 1710, Nipomo, CA 93444
805-929-0887
www.butterflypalace.org

Come to the California Central Coast in the fall and winter and you'll be welcomed by an awesome sight—thousands of Monarchs overwintering in the few remaining microclimates that allow them to rest before migrating to Mexico. Pismo Beach is a popular place for Monarchs to gather but the area is rapidly falling victim to development that is uprooting the Eucalyptus and Monterey Pine groves that Monarchs need to survive.

One woman, Sheila Boone, a fifth great-granddaughter of famed American adventurer Daniel Boone, is hoping to save the Monarch and build a monumental butterfly house while she's at it. Boone has been steadily raising Monarch awareness and funds for a crystal butterfly palace to be constructed in Nipomo, a community just outside of Pismo Beach.

The domed butterfly conservatory, patterned after the historical Crystal Palace in Britain, will feature live tropical species. The conservatory is expected to be just one component of the **Butterfly Palace**. An **American Western Monarch and Rare Butterfly Education and Research Center** will focus on endangered habitats like the Monarch's and the world's rarest butterfly, the Palos Verdes Blue, a species that lives in a small section of Los Angeles. Several museums and 150 acres of showcase theme gardens are planned as well. But it all begins with the live butterfly conservatory.

Boone has launched a major effort to raise money for the non-profit project, so be sure to contact her if you can help.

Las Palmas Park Butterfly Garden (proposed)

National City, CA
doug@basiclink.com
http://sdbirds.basiclink.com/las_palmas_park_butterfly_garden.htm

Help turn languishing parklands into a beautiful butterfly park. Douglas Aguillard, publisher of *SoCal Field Guides*, has initiated an effort to transform a one-acre park parcel into a butterfly haven. The park area has a lovely garden patch with a circular sidewalk, an irrigation system and parking. Litter and graffiti will be removed by the city and volunteers are needs to help plant and maintain the grounds. If the project is as successful as Aguillard's own nearby butterfly-attracting backyard, dozens of native butterflies should visit.

Colorado

Butterfly Pavilion & Insect Center

6252 West 104th Ave., Westiminster, CO 80020
303-469-5441
www.butterflies.org

With the snow topped Rockies looming nearby, it seems incredible that a warm tropical rain forest lives under the shadow of these mammoth mountains. But just take a walk along West 104th Avenue and you'll spot the 16,000-square-foot home of tropical butterflies as well as exotic insects.

Come in from the cold and bask in the warm and humid conservatory where Green Banded Peacocks, Checkered Lime Swallowtails and Montezuma Cattle Hearts swirl around you. These are just three of the more than 50 species you might find flitting among the 100 different species of tropical and sub-tropical plants such as Egyptian Starflower, Chinese Hibiscus and Lantana. About 1,200 flying flowers flutter about at any one time including Blue Glassy Tigers, Blue Striped Crows, Blue Pansies, Paper Kites, Plain Tigers and Red Crackers. Look for a colorful butterfly identification guide on the pavilion's Web site.

Newborn scale wings such as Sailors, Clippers, Queens and Monarchs are released every day into the pavilion. You can watch the releases in the **Chrysalis Viewing Area** at 12:30 p.m. and 3:30 p.m. daily. The **Conservatory** is also home to 22 tortoises, 4 turtles, 2 Ring-Neck Doves, frogs, fish, land crabs, and a very large but very friendly iguana.

Adjacent to the conservatory a section dubbed the **Crawl-A-See-'Em** is devoted to insects and invertebrates. You get a 360-degree view of the miniature habitats housing Giant Sonoran Centipedes, Madagascar Hissing Cockroaches, Texas Brown Tarantulas, Giant African Millipedes, Pink Winged Walking Sticks, Jungle Nymphs, Giant Black Stag Beetles and White-Eyed Assassin Bugs.

Aquatic invertebrates swim in the **Water's Edge** exhibit. Look for Chocolate Chip Stars, Leather Sea Stars, Horseshoe Crabs and Atlantic Pencil Urchins.

Located on five acres, the facility already has plans to spread its wings. The habitat is so popular that the pavilion has to turn away 20,000 school children every year so it will open a 40,000-square-foot glass pyramid building across the street. When completed it will be the world's biggest butterfly house and insect zoo.

In the meantime, you can still see the wild wonders at the original pavilion and center. While waiting for its new pavilion to emerge, the **Butterfly Pavilion** has temporarily expanded its educational space by erecting an onsite Tipi to serve as a classroom.

Hours: 9 a.m. to 5 p.m. daily, until 6 p.m. Memorial Day to Labor Day. Closed Thanksgiving and Christmas Day.

Admission: $6.95 adults, $4.95 seniors (62+), $3.95 ages 4–12, free for ages 3 and under.

Tropical Discovery

Denver Zoo

2300 Steele St., Denver, CO 80205
303-376-4800
www.denverzoo.org

Topped by two pyramid-shaped glass domes, this exhibit is worth the 11 years it took to create. Enter this lush indoor tropical haven and the sights, sounds and smells of a tropical rain forest surround you. The habitat is anything but the ordinary. A journey along the winding pathway takes you to commanding waterfalls, a darkened cave, temple ruins in a jungle, a tropical river bank, a Mangrove swamp, coral reefs, a Cypress swamp and a tropical marsh.

More than 1,200 animals call this place home and they are as varied as the tropical plants. Imagine meeting everything from Poison Dart Frogs to Vampire Bats to Clouded Leopards. And yes, along the way you'll meet some insects and other invertebrates. The first one you'll encounter is the Flamboyant Flower Beetle housed in a glass-fronted cameo exhibit near the entrance. Wind past the Malayan Tri-Colored Squirrels, Tokay Geckos and King Cobras and exit the temple ruin. You'll find some of the spineless wonders you've been looking for in the **Tropical Biodiversity** alcove: Emperor Scorpion, Madagascar Hissing Cockroaches and a variety of tarantulas such as the Goliath Bird Eater. In the **Mangrove Swamp** is a colony of Fiddler Crabs. At the **Tropical Reefs**, the centerpiece is the 15,300-gallon coral reef tank with 80 specimens representing 20 to 25 species. Other aquatic invertebrates include Giant Clams, Coconut Crabs, Sea Anemones, Turbo Snails, Nautilus and Sunburst Diving Beetles. At the end of the center is the world's largest indoor Komodo Dragon habitat. When you exit through the **Discovery Center** you'll find educational exhibits about the animals you've just seen and an animal handler may bring out some critters to see up close.

Hours: 9 a.m. to 6 p.m., April 1 to September 30; 10 a.m. to 5 p.m., October 1 to March 31.

Admission: $9 adults, $7 seniors (62+), $5 ages 4–12, free for ages 3 and under April to September; $7 adults, $6 seniors, $4 ages 4–12, free for ages 3 and under October to March.

Western Colorado Botanical Gardens & Butterfly House

641 Struthers Ave., Grand Junction, CO 81501

970-245-3288

www.wcbotanic.org

When you approach this 12.6-acre blossoming landscape, two 10-foot-tall butterfly sculptures forged by artist-blacksmith Bill Stoddard beckon you to come in. It's hard to imagine that this area near the Colorado River was once a salvage yard. Members of the **Western Colorado Botanical Society** and other volunteers spent years clearing the land for outdoor gardens, a greenhouse filled with orchids and 600 varieties of tropical plants and a small but inviting butterfly house. Flying in the 1,500-square-foot glass-enclosed greenhouse are about 10 species of Native American butterflies from the Gulf Coast: Fritillaries, Queens, Julias, Monarchs, Tiger Swallowtails, Spicebush Swallowtails, Palamedes and Malachites. The facility gets about 100 new pupae every week so during any visit you may see 150 to 200 winged wonders fluttering about. A circular walkway lets you stroll around the greenhouse and you can rest on the path's benches. Inside the walkway are nectar-bearing plants and a goldfish pond. The **Puparium** gives you a close up view of the 50 to 100 butterflies that emerge each week.

Hours: 10 a.m. to 5 p.m. Tue–Sun, April 1 to October 30; 10 a.m. to 4:30 p.m., November 1 to March 30.

Admission: $3 adults, $2 seniors and students with I.D., $1.50 ages 5–12 years, free for ages 4 and under.

Connecticut

UTC Wildlife Sanctuary

Science Center of Connecticut

950 Trout Brook Dr., West Hartford, CT 06119
860-231-2824
www.sciencecenterct.org

Home to more than 50 species of reptiles, birds and mammals, most of them rescued from not so happy fates, the sanctuary has a few invertebrates on display. Built into the wall with glass fronts are the homes of a Giant African Black Millipede, colonies of Madagascar Hissing Cockroaches and South American Cockroaches, a handful of Emperor Scorpions, a host of Giant Prickly Stick Insects and a pair of Chilean Rose Hair Tarantulas. Not on display is a lonely Black Widow who sits in the office of Animal Curator/Educator Cindy Dulac. Found in a shipment of grapes at a local supermarket, you might be able to take a peek if you ask.

Although there are no hands-on activities with the bugs, school groups can book an insect class. The center also has several reptiles such as iguanas, Boa Constrictors, geckos and a Bearded Dragon. Other wild animals you might encounter: a Jungle Cat, woodchuck, lynx, caracal, owls and Prairie Dogs. Don't forget to find **Turtle Town** featuring 11 turtles and tortoises.

Hours: 10 a.m. to 5 p.m. Tue.–Sat., Noon to 5 p.m. Sun. Closed Mondays, Thanksgiving, Christmas and Easter. Open Mondays during holidays and West Hartford school vacation days.

Admission: $6 adults, $5 seniors (65+) and ages 3–12, free for ages under 3.

Roaring Brook Nature Center

70 Gracey Rd., Canton, CT 06019
860-693-0263
www.sciencecenterct.org

About 12.5 miles from the **Science Center of Connecticut** is a satellite site that serves as a nature center. Although not large, the 500-square-foot butterfly garden in front of the center captures the attention of about 20 species of native butterflies. Look for Fritillaries, Angle Wings and Skippers among the Purple Coneflowers, Milkweed and Sassafras. The nature center offers butterfly programs during the summer, including one for pre-schoolers. Be sure to find out about the center's guided nature walks and directions for taking the trails through **Werner's Woods**.

Hours: 10 a.m. to 5 p.m. Tue.–Sat., 1 p.m. to 5 p.m. Sun. Open Mondays in July, August and on Holidays.

Admission: Free.

Connecticut State Museum of Natural History

University of Connecticut

2019 Hillside Rd., Unit 1023, Storrs, CT 06269
860-486-4460
www.mnh.ucon.edu

Under extensive renovation, this small natural history museum is big when it comes to events. It sponsors more than 200 programs throughout the year including several on butterflies and butterfly gardening. The biggest event is **BioBlitz**, a 24-hour biological survey of life in a Connecticut park. With a base camp set up with microscopes and collecting kits, scientists scope out a different park each year. The public is always invited to share in the discovery and in 2001, 168 scientists recorded 2,519 species including dozens of bugs and butterflies. Part contest, part festival, part scientific adventure, **BioBlitz** celebrates life in your own backyard. The event is held annually, but not necessarily at the same time each year, so be sure to check with the museum for time and place.

Despite the renovation, the museum still maintains a small exhibit space. From April 15 to July 2002 that space spins with the excitement of **Butterflies**, a temporary display that explores the relationships between people and the environment. Thousands of butterfly specimens from the museum's collections will be on view and several hands-on activities are planned for visitors of all ages.

Since there's always something new including a program or event involving bugs or butterflies, be sure to visit the museum if you're stopping by the university. The campus also has a **Co-op Bookstore** with nature-themed books and toys; and animals barns where you can see horses, sheep and cows. Sample **UConn**'s prize-winning ice cream or picnic on Horsebarn Hill or on the lawn around Mirror Lake.

Hours: 10 a.m. to 4 p.m. Mon.–Fri. Open occasional weekends; closed major holidays.

Admission: Free, donations appreciated.

Peabody Museum of Natural History

Yale University

170 Whitney Ave., New Haven, CT 06511
203-432-5050
www.peabody.yale.edu

Visit the place where insects are turned into tasty desserts for the annual **Edible Insects** event. Although the museum does not have any live insects on display it does exhibit a portion of its one million specimens. Look for the beetle display as you walk toward the **Discovery Room** on the second floor. An ornate work of art uses beetles to form the shape of a Scarab Beetle. Insects pop up at odds and ends of displays and about half of the museum's dioramas sport some bugs.

One of the best times to visit is during the **Edible Insects** insect tasting event in October. You'll also find live bugs there—not meant for consumption—plus several other buggy activities.

Hours: 10 a.m. to 5 p.m. Mon.–Sat., Noon to 5 p.m. Sun. Closed major holidays.

Admission: $5 adults, $3 seniors (65+) and ages 3–15.

Washington, D.C.

Otto Orkin Insect Zoo

Smithsonian National Museum of Natural History

10th St. and Constitution Ave., NW, Washington, D.C. 20560
202-357-2700
www.mnh.si.edu/museum/VirtualTour/Tour/Second/InsectZoo/

Welcome to the granddaddy of all insect zoos and the home of Cleo, Muriel and Miss Piggy, three of the zoo's oldest tarantulas. Opened year round since 1976, the zoo was the first permanent exhibit of live insects and other invertebrates any where in the Americas. Since it's located inside the **Smithsonian**, it's probably the most visited insect zoo in the world, drawing more than one million visitors a year.

Bug-eyed Lubber Grasshoppers and wagging Honey Bees leap and buzz under the direction of a passionate bug man, entomologist Nate Erwin. **Insect Zoo** director since 1995, Erwin coos over cockroaches and feeds the voracious Leafcutter Ants which farm fungus for a living. He welcomes visitors to ask questions, to learn about the science of entomology and to understand that bugs need love too.

There are plenty of bugs to love at the zoo: Australian Stick Insects,

Black Widow Spiders, Bombardier Beetles, centipedes, German Cockroaches, bright black and orange Milkweed Bugs, Praying Mantids, scorpions, Stink Bugs, New Guinea Walking Sticks, Silk Worms, Silverfish, katydids, Jumping Bristletails and Velvet Ants.

Visitors can interact with the bugs, too. Mexican Red Knee Tarantulas feast on crickets during daily live feeding demonstrations in the **Southwest Desert Diorama** area. From Tuesday to Friday, mealtimes are at 10:30 a.m., 11:30 a.m. and 1:30 p.m. and on Saturday and Sundays, the Big Ts get snacks at 11:30 a.m., 12:30 p.m. and 1:30 p.m. On some days, volunteers may be able to offer insects to pet such as a Madagascar Hissing Cockroach or a Tobacco Horn Worm (really a butterfly caterpillar).

Watch Honey Bees make their golden nectar in a **Bee Tree** exhibit then turn your eyes to an adjacent window where you can see inside the insect-rearing lab. Discover how Erwin and his staff keep the insect zoo alive with creeping and crawling creatures.

Visit the **Termites' Turf**, where kids can crawl through a 14' model of an African Termite mound. In real life, as many as 2 million termites would be streaming through the mound tunnels but in this exhibit you thankfully get to watch footage of termite activities on two video monitors.

Want to know what bugs could be crawling in your own home? Then take a look inside a replica of a house, from an insect's point of view. The model family home shows how insects are quite cozy on furniture, clothes, wood and family pets. Look at live silverfish, Flour Beetles and German Cockroaches and learn about other common insects such as Clothes Moths, Carpenter Ants, fireflies, fleas and mosquitoes with push buttons on colorful insect cases.

Go beyond your familiar surroundings into the museum's tropical **Rain Forest** exhibit and explore how biodiversity maintains the balance in the rain forest ecosystem. You'll find live Giant Madagascar Hissing Cockroaches, a Leaf Cutter Ant colony, and several cave arthropods such as Amblypigids, Tail-Less Whip Scorpions and Peripatus, a creature considered a

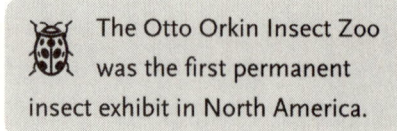

> The Otto Orkin Insect Zoo was the first permanent insect exhibit in North America.

link between insects and worms.

After hearing about all these live bugs, you may wonder why this **Smithsonian** institution is named after the founder of a bug exterminating company. Fact is, the zoo underwent a wonderful metamorphosis in 1993 after Orkin donated half a million dollars to renovate the aging facility. So the zoo was named after another granddaddy, Otto Orkin (1887–1968). Orkin continues its relationship with the zoo through **Insect Safari**, an insect zoo on wheels that visits more than 50 places annually (see Bugged Out chapter).

Hours: 10:00 a.m. to 5:30 p.m. daily.
Admission: Free.

Pollinarium

National Zoo

3001 Connecticut Ave., N.W., Washington, D.C. 20008
202-673-4950
www.natzoo.si.edu

Washington is all abuzz and it isn't just about the latest political gossip. Bees are creating quite a stir at the **National Zoo**. The bees, along with butterflies, birds and other pollinators, are on display at this free 163-acre zoological "BioPark" that illustrates the important relationships among plants, animals and humans.

Bright foliage punctuates a lush garden that's at the heart of the 1,250-square-foot glass-enclosed **Pollinarium**. Hundreds of Zebra Longwing Butterflies float among the blossoms of Orchids, Lantana and Passionflowers that also entice the free-flying Anna's Hummingbirds. A banquet of butterfly species zip through the flowering flora of Fire Bush, Egyptian Star Flowers and Jungle Flame: Cloudless Sulphurs, Queens, White Peacocks, Gulf Fritillaries, Orange-Barred Sulphurs and Julia Butterflies.

Step inside a replica of a seven-foot-tall hollow tree and safely watch as thousands of Italian Honey Bees work to make honey. The exhibit gives you the sense of actually entering inside the hive as you observe the bees buzzily building combs and tending to the young bees. You'll see the bees buzz in and out of an acrylic tube that let's them go outside the exhibit to forage for pollen and nectar.

Illustrated educational displays explain the importance of pollinators. One-third of all the plants eaten or used for fiber in the U.S. need pollinators to grow. Look for the six-foot-long model of a Sage Flower with a four-foot Honey Bee. Artistically designed to be 100 times larger than life, you can move the bee so it appears to take pollen from the anatomically correct flower. There's also a life-sized re-creation of a blooming Saguaro Cactus and a pollinating bat.

> Swarming bees usually don't sting.

Located behind the **Reptile Discovery Center**, you enter the **Pollinarium** through a smaller **Invertebrate Exhibit**. The entrance is just past the Orb Weaver Spiders. The exhibit features corals, Sea Anemones, Spiny Lobsters, octopuses, Madagascar Hissing Cockroaches and other spineless wonders, just a portion of the 5,800 animals that call the zoo home.

Hours: (Grounds) 6 a.m. to 8 p.m. (Buildings) 10 a.m. to 6 p.m., May 1 to September 15. (Grounds) 6 a.m. to 6 p.m. (Buildings) 10 a.m.–4:30 p.m., September 16 to April 30. Closed Christmas Day.

Admission: Free.

Butterfly Habitat Garden

Smithsonian National Museum of Natural History

10th St. and Constitution Ave., NW, Washington, D.C. 20560
202-357-2700
www.mnh.si.edu

Butterflies blossom at the first outdoor educational experience created by the world's largest natural history museum. At least 30 species of butterflies have been recorded at this 300' by 40' butterfly garden that depicts four distinct types of butterfly habitats: **Wetlands**, **Meadow**, **The Wood's Edge** and **The Urban Garden**. Planted in 1995, it's the first garden to bloom on the east side of the museum, near the Ninth Street Tunnel. Descriptive signs help you identify the plants and butterflies native to each habitat. More than 2,500 herbaceous plants and 200 woody shrubs and trees provide nectar while Cabbages, Fennel and Parsley woven among the ornamental foliage provides a place for butterflies to lay eggs and food for caterpillars.

Zebra Swallowtails and Viceroys favor the **Wetlands** portion of the garden, which mimics wet habitats found on the East Coast near marshes, bogs and ponds. Plants that can handle the excessive moisture provide a place for butterflies to puddle or use their proboscis (long tongue) to sip water.

Alfalfa Butterflies and Eastern Tailed Blues find the dry **Meadow** habitat welcoming. With open spaces filled with wildflowers and grasses, this habitat usually found in the Midwest and Central U.S. is perfect for butterflies. The open area lets sunshine pour in and butterflies develop faster from larvae to adults.

On the **Wood's Edge**, Tiger Swallowtails and Red Admirals flourish. A transition zone between the meadows and the **Urban Garden**, you'll find sheltering plants, trees and shrubs that allow butterflies to overwinter and provide save spaces for larvae to survive.

Discover how you can attract Black Swallowtails, Cabbage Whites and Silver Spotted Skippers to your own garden as you watch these flitting flowers float among plants that typify a butterfly-friendly urban garden. Most of the plants rooted here are readily available at nurseries, native plant societies and garden catalogs.

Hours: Open all day, year round.
Admission: Free.

Delaware

Ashland Nature Center's Butterfly House (reopens 2003)

Delaware Nature Society

Junction of Brackenville and Barley Mill Roads
P.O. Box 700, Hockessin, DE 19707
302-239-2334
www.delawarenaturesociety.org

Amid the meadow, marsh and forest of this 600-acre nature center is an educational greenhouse that may be the only facility in the state that features live butterflies. The 18' by 30' mesh enclosure is strictly a natural butterfly house providing food and shelter for both adult butterflies and their caterpillars. The shade cloth house is open to the elements, enticing about 20 species of butterflies to fly in. You can observe the complete life cycle from egg to emerging adult: discover tiny Monarch eggs under a Milkweed leaf, a Painted Lady caterpillar munching on Thistle and a Swallowtail chrysalis hanging from a Spicebush twig.

In mid-July you'll see up to 100 butterflies winging through the house: Monarchs, Black Swallowtails, Great Spangled Fritillaries and Commas. You might also see moths such as Giant Silk Moths, Cecropias and Lunas. Take a guided tour with a nature society naturalist and learn about metamorphosis, the food web and the importance of maintaining our natural habitats. The tour includes the butterfly house, a walk through the nearby **Native Plant Demonstration Garden** and an exploration of butterfly habitats outside the house.

Ask about the self-guided trails through the center's various natural habitats and be sure to examine the educational exhibits and visit the **Children's Discovery Corner.** If you live nearby, consider becoming a member to take advantage of the special members-only programs. You'll get to visit some of the nature society's limited access preserves that offer a bounty of butterflies, birds and wildflowers.

Hours: (Butterfly House) June to August. Closed in 2002 due to construction. Opens 2003.

Admission: $2 adults, $1 ages 3–12.

Florida

Butterfly World

Tradewinds Park

3600 W. Sample Rd., Coconut Creek, FL 33073
954-977-4400
www.butterflyworld.com

Walk into the queen of all butterfly parks and you're truly in the middle of a lush tropical paradise, sprayed by mist and fanned by the delicate wings of thousands of butterflies and birds. Nearly three acres in size, **Butterfly World** was the first of its kind when it was built in 1988 and it remains the largest live butterfly park in the world.

Comprised of several butterfly, bird and garden habitats, the **Tropical Rain Forest Aviary** is the crown jewel of the park. A kaleidoscope of more than 5,000 butterflies whirl and wave through this 8,000-square-foot screened yet open-air environment. Up to 80 butterfly species from South and Central America, Malaysia, Taiwan and the Philippines flutter through a butterfly heaven.

Look for species not often seen at other butterfly houses: Tailed Jays, Peacock Pansies, Emerald Swallowtails, Rusty Tipped Pages, Red Rims,

Blue Glassy Wings, Striped Blue Crows, Fatimas, Cruisers, a variety of Piano Keys, the purple Archduke and golden Autumn Leafs. There's also the popular electric Blue Morpho and its cousin the Banded Morpho, Clippers, Plain Tigers, Lacewings, Limes, Owls, Coolies, Cattlehearts, Common Roses, Rosinas, Rice Papers, Saras and Malachites. At various times you may also see some unusual moths such as the Luna, Atlas and Composia.

As you walk through the tropics, rainbows of flowers and trees are in abundance and a light shower refreshes you and the butterflies every five minutes. Explore the pathways leading to a waterfall and small cave, or feed the Japanese Koi in the ponds.

Outside, you'll see thousands more butterflies of the local variety busily darting among the richly planted butterfly-friendly gardens. Among them: Monarchs, Atalas, Cassius Blues, Ruddy Daggerwings, Orange-barred and Cloudless Sulphurs, White Peacocks, Zebra Longwings, Queens, Gulf Fritillaries, Julias, Giant and Gold Rim Swallowtails.

When you first enter **Butterfly World** you'll be offered self-guided information in English, Spanish, French or German. Your first stop should be the **Butterfly Farm** where you can see the entire butterfly life cycle through laboratory windows. Most of the butterflies released here are raised on this farm.

Next is the **Paradise Adventure Aviary** where thousands of butterflies from five continents skip along the flower-lined pathways. This particular flight house is home to thousands of plants that butterflies lay their eggs on. The plants are clearly marked so you can see this amazing work of nature.

Tucked between this aviary and the **Tropical Rain Forest** is the **Hanging Garden** and **Butterfly Emerging Area.** Flowing with fragrant hanging baskets that provide nectar for butterflies, the garden also features mesmerizing "sensitive plants." The plants fold their leaves and appear to wilt at the slightest touch. Glass cases filled with chrysalides line one wall and you can watch as butterflies emerge and spread their wings to dry.

Be sure to visit the **Secret Garden**, a maze of one of the largest Passion Vines collections in the world. Look above the canopy of Dutchman's Pipe blossoms and Sky Vine lavender flowers to see native butterflies weaving through the mass of foliage. Also visit the museum and insectarium with specimens of butterflies, moths and other insects from around the world. The **Butterfly Gardening Plant Shop** offers a bit of paradise to take home in the form of native plants that will attract winged wonders to your own backyard. Stop at the **Butterfly Café**, an outdoor eatery where you can snack and enjoy the surrounding wildlife. The **Caterpillar Canopy Gift Shop** is filled with butterfly gifts and books.

Hours: 9 a.m. to 5 p.m. daily.

Admission: $12.95 adults and seniors, $7.95 ages 4–12, free for ages 3 and under.

Wings of Wonder, The Butterfly Conservatory

Cypress Gardens

2641 S. Lake Summit Dr., Winter Haven, Fl 33884
800-282-2123
www.cypressgardens.com

Enter the 5,000-square-foot glass-enclosed Victorian-style conservatory—the only one of its kind in Florida—and you enter a tropical paradise. About 1,000 jewel-toned flying wings zip around the palms, ornamental figs and flowing flowers. About 50 species of tropical and sub-tropical butterflies hail from exotic countries such as Kenya, Surinam, Ecuador and Belize. Walk among iridescent Blue Morphos, Rice Papers, Owls and Small Postman Butterflies. Fluttering along with the exotics is the Zebra Longwing, Florida's state butterfly.

Butterflies aren't the only creatures you'll encounter flying free. The conservatory is home to a host of colorful exotic birds including the very rare Bleeding Heart Dove from the Philippines and Button Quail from Southeast Asia. A large pool and waterfall are located in the center of this butterfly bliss. Mounted chrysalis cases filled with 600 to 1,000 pupae let you view emergence up close. Adjacent to the indoor habitat are butterfly and herb gardens where you'll discover several native butterflies flying.

Florida's first theme park, the 200-acre **Cypress Gardens** celebrated its 65th anniversary in 2001 and features other animal habitats like **Gator Gulch** stocked with 50 reptiles including Mighty Mike, a 13.5' alligator. Look for the **Birdwalk Aviary** with colorful Lorikeets that eat out of your hand, a southern-style zoo, and **Nature's Arena**, a live animal show. And those are just the animals! Located on the shores of Lake Eloise, the park features world-famous water ski shows, botanical boat tours, and cruises on the **Southern Breeze Paddle Wheeler**.

Hours: 9:30 a.m. to 5 p.m. daily; extended summer and holiday hours.
Admission: $34.95 adults, $19.95 ages 6–17, free for ages 5 and under.

The Butterfly Sanctuary

Naples, FL
941-354-1098
www.thebutterflysanctuary.com

Welcome to butterfly heaven. Nestled in a natural haven in a corner of Southwest Florida that borders the Everglades, this is indeed a sanctuary for species of flutterbys trying to escape the encroaching development in Naples. For Mia Mazza and her artistic family of five, the sanctuary is home. Mia is a realistic nature artist, Tropical Tommy a local musician, daughter Jackie makes crafts, daughter Julie is an artist and son Jason, a writer. All have worked since 1999 to create what they call a refugee camp for butterflies.

The family has cultivated their patch of paradise with a focus on the Florida wetlands as well as Florida's dry scrub. Species that visit the sanctuary are monitored and the family keeps lists of the butterflies, moths and other wildlife they observe. To date 33 species of butterflies and 11 species of day flying moths have visited. Colonies of Monarchs, Queens and Sulphurs overwinter here. The Zebra Longwings have a colony with a great-great-great-grandmother who decides when the group rests and wakes up.

During your visit you may see clouds of native butterflies fluttering in the open. The nature trail features 60 different larval plants and dozens of nectar plants that bloom year-round because of the tropical type of climate. A new edition that lets you get even closer to these winged beauties is a small 12' by 24' by 12' high butterfly house. The **Flight House** is brimming with host and nectar plants that attract local species. You can see several different caterpillars munching and sometimes observe the magic of a butterfly emerging from a pupa.

The family offers tours to visitors but you must make an appointment. The **Butterfly Sanctuary** address is not published but you'll get directions when you book your tour. The sanctuary also conducts butterfly gardening classes on Saturday mornings at 9 a.m. Once you visit you might want to join this nature conscious clan by reseeding roadsides with native wildflowers, joining the **Butterfly Club** or participating in its ongoing **Butterfly Count**.

Be sure to stop in Mia's small art studio which features some educational materials, butterfly toys, books Mia has written and illustrated about butterflies, her artwork and painted wooden crafts.

Hours: By appointment. Butterfly gardening classes at 9 a.m. Saturdays.

Admission: Donation (goes to maintaining the trail and wildlife). $10 or donation for butterfly gardening class.

Upward Bound Butterfly Garden

Miami Museum of Science

3280 South Miami Ave., Miami, FL 33129
305-646-4200
www.miamisci.org

Because of a group of caring teens, caterpillars turn into butterflies at the **Miami Museum of Science**. The crew of 35 or so members of the inspirational organization Upward Bound transformed an 84' long plot of earth outside the museum's **Wildlife Center** into a native butterfly habitat. A dozen of the kids, with help from Upward Bound director Jennifer Schooley and museum staff, designed and planned the project in 1999 while 25 students helped construct and plant the outdoor butterfly haven.

Now visitors to the museum can watch in wonder as Giant Swallowtails, Cloudless Sulphurs and Gulf Fritillaries fly through the garden. A red brick walkway, trellises overflowing with flowering vines and a streaming fountain give the garden a sanctuary feel. The garden is filled with a mix of native Florida and exotic butterfly host and nectar plants: Coontie for the Florida Atala Butterfly, Dill for the Swallowtails, Milkweeds for the Monarchs and Cassia for the Sulphurs. Educational signs describe the foliage and the butterfly species. Don't forget to visit the adjacent **Wildlife Center** with 175 different birds of prey and reptiles.

Hours: 10 a.m. to 6 p.m. daily. Closed Thanksgiving and Christmas.

Admission: $10 adults, $8 seniors (62+) and students with I.D., $6 ages 3–12, free for ages under 3.

The Habitat

Gulfcoast Wonder and Imagination Zone (G.Wiz)

1001 Boulevard of the Arts, Sarasota, FL 34236
941-906-1851
www.gwiz.org

Visit a slice of wild Florida when you enter this climate-controlled exhibit of butterflies and other native Florida flora and fauna. Filled with brilliant flowering plants such as Pentas, Lantana and Butterfly Bush, the two-story glass enclosure offers a spectacular view of 400 delicate wings of color. About a dozen species can be spotted including Gulf Fritillaries, Julias, Viceroys, White Peacocks, Giant Swallowtails and in line with Florida's citrus history, Sleepy Oranges. You can even see them fluttering from a second story viewing area. At the ground level you can watch all the stages of butterfly life. Look for eggs and caterpillars on host plants like Tropical Milkweed and Cassia. Find the pupas in an emergence box then watch as butterflies break free from their chrysalides then unfold and dry their wings.

The butterflies aren't the only living creatures in this 80-degree year-round habitat. An observation Honey Bee hive buzzes with life as you watch worker bees make wax and the Queen lay her eggs. Green Tree Frogs, native to Florida and the introduced Cuban Tree Frog live in animal enclosures, as do Mr. Moo, a Red Rat Snake and Maizy, a part albino Rat Snake. The snake duo is brought out for close encounters and you can have your photo taken with them. Aquatic live-in residents include Red-Eared Slider Turtles, Tilapia, Blue Gill, Mosquito Fish, Catfish and an Alligator Gar.

Hours: 10 a.m. to 5 p.m. Tue.–Sat., 1 p.m. to 5 p.m. Sun.

Admission: $7 adults, $6 seniors, $5 ages 2 and older.

BioWorks

Museum of Science & Industry (MOSI)

4801 E. Fowler Ave., Tampa, FL 33617
813-987-6100
www.mosi.org

Outside the largest science center in the Southeastern U.S. is a 6,400-square-foot Florida-friendly butterfly garden. Created in conjunction with the Southwest Florida Water Management District, **BioWorks** demonstrates how water-conserving technology can still produce a garden blooming with blossoms and butterflies.

Discover how "the land of flowers" has a unique ecosystem and residents can use the right plants to create a butterfly buffet in their own backyard. Located southeast of the museum's main entrance, the garden is vibrant with the colors of 200 plants from Beach Sunflowers to Firebush to Blazing Stars. Brilliant too are the butterflies, about 175 of them on any given day, representing six to nine local species.

There are five hands-on activities too: a microbiology station with microscope viewing; biology station and nutrient puzzle; a botany station with magnifying glasses; an engineering station with sediment settling jars; and a chemistry station for pH measuring. Check the science center's schedule for special butterfly presentations in the **Main Entrance Plaza** just north of **BioWorks**.

Afterwards, head south to **The Back Woods** outdoor exhibit featuring a large variety of plant and animal life. Look for the sinkhole and populations of protected gopher tortoises among Pine Flat Woods, Turkey Oak Sandhills, Oak Hammocks, and Wetlands. Be sure to pick up the "Ecology to Go" tool boxes containing head sets for an audio tour, binoculars, microscopic viewers, field guides, and identification cards.

Hours: 9 a.m. to 5 p.m. Mon.–Fri., 9 a.m. to 7 p.m. Sat.–Sun. Open 365 days a year, reduced hours on certain holidays.

Admission: $13.95 adults, $11.95 seniors (60+), $9.95 ages 2–12, free for ages 2 and under.

Harrell Discovery Center

Lowry Park Zoo

1101 West Sligh Ave., Tampa, FL 33604
813-935-8552
www.lowryparkzoo.com

Considered one of the top three mid-sized zoos in the U.S. and Tampa's least expensive science-education experience, this 41-acre animal adventure has a handful of creepy crawlies to show off. Residing in a small 1,500-square-foot interactive area are Mexican Red Knee, King Baboon, Cobalt Blue and Pink Toe Tarantulas. There's also a colony of Brazilian Cockroaches and an Emperor Scorpion. The center also has a variety of reptiles and unusual amphibians such as Poison Dart Frogs, African Bullfrogs and Asian Leaf Frogs—they really look like a leaf from the top!

Located inside a 100-acre city-owned park, the zoo opens a brand new children's area called **Wallaroo Station** in June 2002. The 4.5-acre Australia themed mini-zoo offers families some playful and safe animal encounters.

Hours: 9:30 a.m. to 5 p.m. daily. Closed Thanksgiving and Christmas Day.

Admission: $9.50 adults, $8.50 seniors (50+), $5.95 ages 3–11, free for ages 2 and under.

Panhandle Butterfly House

The Nature Walk

Navarre Park, Highway 98, Navarre, FL 32566
800-480-7263 or 850-939-3267
www.angelfire.com/nb/nbh/index.html

You can't miss the colorful set of wooden wings that announce the entrance to this butterfly house kept open and maintained by a cadre of more than 100 passionate volunteers. It graces a section of a park aptly named **The Nature Walk** for a one-mile stretch of path that parallels the shores of Navarre Beach. Skirting the Florida Panhandle's Emerald Coast, the area is perfect for a house full of flying native gems.

Now an educational facility run by a consortium of Master Gardener associations from Santa Rosa, Escambia, and Okaloosa counties, the house was originally a tourist attraction. Inside the misty 60' by 40' Quonset hut you're met by waves of flowering plants and light clouds of fluttering Florida butterflies. A palette of more than 200 butterflies representing a dozen or so different species can be seen on any given day. Among them, Zebra Longwings (the official state butterfly), Gulf Fritillaries, Painted Ladies, Buckeyes, Red Admirals, Viceroys, Monarchs and Queens. Swallowtails are popular with Tiger, Spicebush, Giant and Palamedes varieties and there are several types of Sulphurs and Hairstreaks. Unlike most butterfly houses, the majority of the butterflies here are purchased as live adults from a Florida butterfly farm.

The Master Gardeners have created a showcase for gardens that are a literally banquet for local butterflies with Verbena, Sunflowers, Zinnias and Pentas for adults. Even more important, says butterfly house director and Master Gardener Sandra Sherman, are the plants for the caterpillars such as Dill, Parsley, Milkweed and Cassia. Without them, adult butterflies won't spend too much time in your garden because they can't lay eggs. If you see something you like and want to plant it in your garden, the butterfly house probably has it for sale.

Look for the educational nursery area where eggs and caterpillars have been gently removed from the butterfly host plants and placed inside cages with host plants. You can watch as eggs turn into caterpillars, as caterpillars munch voraciously, as caterpillars turn into chrysalides, and as adult butterflies emerge from their casings.

As you enter the enclosure's reception area you can't help but stop and look at the collection of pinned butterflies. A local college professor donated the crop of 450 specimens; compare butterflies from around the world with local ones.

Butterfly friendly gardens also surround the greenhouse and the park is family-friendly with a duck pond, children's play area and picnic tables. While you enjoy lunch you get a perfect view of the Navarre inland waterways and the Gulf of Mexico.

While the butterfly house doesn't charge admission, its does accept donations, and gladly. The house also accepts donations of butterfly attracting plants, butterflies, caterpillars and eggs. Be sure to sign the guestbook and add your name to the visitors from 47 states and 18 countries who have seen this all-volunteer attraction.

Hours: 11 a.m. to 4 p.m. daily, late April through Labor Day.
Admission: Free, donations accepted.

Critters and Things

Sarasota Jungle Gardens

3701 Bay Shore Rd., Sarasota, FL 34234
941-355-5305
www.sarasotajunglegardens.com

Meet showstoppers Sonny & Cher, a pair of Madagascar Hissing Cockroaches when you visit this old-style Florida attraction. The dynamic bug duo often performs in the attraction's **Critters and Things** show in an area near **Reptile World** and the **Alligator and Croc Habitat**.

Featuring the fuzzy, the furry, the creepy, the crawly and everything in between, the animal handlers rotate what creatures they feature. You might see Lubber Grasshoppers, a tarantula, a scorpion or non-invertebrates such as a Gila Monster, a ferret or a hedgehog.

> There are one million different species of insects on Earth.

The jungle's critter, bird and reptile shows are just part of its old-fashioned ambience. Lush gardens, nature trails and plenty of wildlife from Asian Leopards to Gators to Macaws make this 10-acre jungle habitat a true tropical adventure. Look for an outdoor butterfly garden behind the **Flamingo Café**, about a five-minute walk from the main gate.

Hours: 9 a.m. to 5 p.m. daily. Closed Christmas Day.

Admission: $10 adults, $9 seniors (62+), $6 ages 3–12, free for ages under 3.

Greathouse Butterfly Farm

20329 State Rd. 26 East, Earleton, FL 32631
866-475-2088
www.greathousebutterflies.com

Zane Greathouse knows how to tell one pupa from another. And it's a good thing too, because this science teacher turned his family's 100-acre pecan farm into the largest butterfly farm in North America.

The bottom had dropped out of the pecan market several years ago, when Zane, who grew butterflies for his students, came up with the idea of transforming his parents' ailing acreage into a new kind of agriculture. Now this rural parcel that's been in the family since before the Civil War is prosperous again.

From weddings to natural history museums to Hollywood movies, if the butterflies are flying, they just might be from this rural North Florida farm. If you're looking at a Ruddy Daggerwing at the **American Museum of Natural History**, it probably started life as an egg at the farm. Same goes for the Viceroys at the **Panhandle Butterfly House** in Navarre, Florida, or in Oprah's film, *Beloved*.

More than 45 varieties of native Florida butterflies are raised on the farm: American Swallowtails, Giant Swallowtails, Buckeyes, Julias, Great Southern Whites, Malachites, Viceroys, Monarchs, Laurel Swallowtails, Gulf Fritillaries, Orange Barred Sulphurs, Red Spotted Purples, Ruddy Daggerwings, Spicebush Swallowtails, Tiger Swallowtails, Queens, Zebra Longwings and Zebra Swallowtails.

The whole family is involved in the butterfly business. Zane's brother Dan, a retired marine, tends to the plant nursery that supplies the hundreds of nectar and host plants butterflies need to survive. Dan's wife Kay gently scoops up live butterflies to fulfill orders and keeps track of what species are in stock. Butterflies, being seasonal beasts, propagate during certain times of the year. You can find a list of all the species the **Greathouse Farm** produces on their Web site.

Despite being a commercial supplier, the farm also promotes education. After all, this bug biz started with a science teacher's idea. The farm provides butterflies for teachers, freebie presentations and the farm has an exhibit with 1,000 butterflies that travels throughout the Southeastern U.S.

Luckily for Florida tourists, **Greathouse Farm** provides tours of its successful operation during the spring, summer and fall. Tours are offered twice a day three days a week, and you have to book ahead. Enjoy an educational walk through the farm's butterfly gardens and find out which plants attract which butterflies. Tour the special butterfly exhibit where hundreds of airborne rainbows are swirling. You'll also see the butterfly barn and nursery where you'll learn how the farm grows such a beautiful bounty, including eggs, larvae, pupae and adults.

There's a small gift shop with T-shirts, books, butterfly feeders and cold refreshments. If you need more sustenance head for nearby Melrose, a quaint Florida town with places to eat.

If you plan to stay in the area for a while, take advantage of the farm's **Butterfly Gardening** and **Butterfly Farming** classes. Twice a month the farm gives a specialized tour that focuses on creating your own butterfly garden. Once a month, you can learn the secrets of growing your own butterflies for pleasure and profit. Sure beats pecans these days.

Hours: (Tours) 10 a.m. and 1 p.m. Tue., Wed. and Fri., mid-March to early November. (Butterfly Gardening) 1 p.m. second and fourth Saturdays each month. (Butterfly Farming) 1 p.m. third Saturday each month.

Admission: (Regular tours) $5 adults, $2 children. (Gardening) $10 person. (Farming) $20 per person.

Center for Lepidoptera Research
(opening 2004)

Florida Museum of Natural History

SW 34th St. and Hull Rd., Gainesville, FL 32611
352-846-2000
www.flmnh.ufl.edu

In 2004 this new 35,000-square-foot **Lepidoptera Center** will house one of the world's largest and complete butterfly collections, a live butterfly exhibit and research and public education facilities. Part of the University of Florida, the museum's center was given its wings by one of the largest private gifts ever donated for insect research—$4.2 million from William W. and Nadine M. McGuire of Minnesota.

The new building will allow the university to move its astounding collection of one million butterflies from an overburdened space at the **Allyn Museum of Entomology**, a research facility in Sarasota, Florida. The collection represents more than 90 percent of all known butterflies.

Georgia

Cecil B. Day Butterfly Center

Callaway Gardens

P.O. Box 2000, Pine Mountain, GA 31822
800-CALLAWAY
www.callawaygardens.org

One of North America's largest glass-enclosed tropical conservatories displaying live butterflies sits in the middle of a 14,000-acre year-round horticultural display garden 60 minutes southwest of Atlanta, GA. Callaway's shimmering octagonal conservatory houses more than 1,000 free-flying winged rainbows that skip inside its 854 glass panes and flutter around a 12-foot waterfall.

The first butterfly house to combine world-class horticulture and the first in the world to showcase African butterflies, the **Day Butterfly Center** features more than 50 tropical species from Malaysia, the Philippines, Kenya and Costa Rica. Observe brilliant Blue Morphos, striking Emerald Swallowtails also known as Banded Peacocks, Scarlet Swallowtails whose markings are reminiscent of a ladybug, Malay Lacewings, Gray Pansies, Clippers, Mosaics, Citrus Swallowtails and Red Passion Flowers. About a third of the butterflies are raised at the 4.5-acre center, with the rest imported from tropical rain forest butterfly farms.

Winging their way with the butterflies are several birds. Hummingbirds dart among the plants while two colorful South American Macaws—one scarlet and one blue and gold—perch in the conservatory along with a green Amazon Parrot from Central America. Look in the landscape for a pair of Mandarin Ducks, a pair of Wood Ducks and an

Asian Crested Partridge.

There are educational displays plus an orientation theater, which continuously shows the award-winning movie *On Wings of Wonder*, depicting the life of a butterfly. Inside the gift shop you'll find a butterfly bonanza of goodies.

Outside the center are 1.5 acres of butterfly gardens with several nectar plants such as Butterfly Bush, Glossy Abelia and Lantana as well as host plants such as Passion Flower Vine, Milkweed and Sage. The visitors to the gardens include several native species: Cloudless Sulphurs, Variegated Fritillaries, Tiger Swallowtails, Monarchs, Skippers and Painted Ladies.

If you have time, take a close look at some of the artwork in the lobby. Original butterfly watercolors by early 19th century artist Chevalier de Freminville are on display after having been lost for more than 100 years. Hanging on the walls leading to the center are watercolors by Georgia pioneer naturalist John Abbot. Glance up at the intricate copper chandelier crafted by artisan Ivan Bailey and you'll see native Passionflower, a major host plant for many butterflies, while if you look down at the carpet underneath your feet you can see Georgia's state butterfly, the Eastern Tiger Swallowtail, woven into the large octagonal rug.

Hours: 9 a.m. to 5 p.m. daily.

Admission: $12 adults, $6 ages 6–12, free for ages under 6. Includes most garden attractions.

World of Reptiles

Zoo Atlanta

Grant Park, 800 Cherokee Ave. SE, Atlanta, GA 30315
404-624-5600
www.zooatlanta.org

Don't let the word "reptiles" fool you when you visit the 40-acre zoo's **World of Reptiles** exhibit. Despite being one of the finest reptile collections in the world with some 400 individual animals, the exhibit is also home to creatures without backbones—invertebrates.

Zebra Longwings are the stars of the zoo's invertebrate collections. In a special glass display that looks like a typical backyard, visitors can watch the "Methuselahs" of butterflies zip through life at a slower pace than other butterflies. While other species live only a couple of weeks, Zebra Longwings live from three to five months because of their ability to digest pollen.

The exhibit lets you view the complete life cycle of the butterflies from eggs to caterpillars to pupae to adult. The eggs, however, are quickly whisked off the plants since the turtles that live with the butterflies eat the eggs.

Sprinkled throughout the **World of Reptiles** are several other bugs: millipedes, Flat Rock Scorpions, Emperor Scorpions, Australian Walking Sticks, a Green Bottle Blue Tarantula, a Goliath Bird Eater Tarantula and Madagascar Hissing Cockroaches.

Hours: 9:30 a.m. to 4:30 a.m. daily. Open until 5:30 p.m. weekends during Daylight Savings. Closed Thanksgiving, Christmas and New Year's Day.

Admission: $16 adults, $12 seniors (55+), $11 ages 3–11, free for ages under 3.

Illinois

Judy Istock Butterfly Haven

Peggy Notebaert Nature Museum

Chicago Academy of Sciences, 2430 North Cannon Dr.,
Chicago, IL 60614
773-755-5100
www.chias.org

A year-round butterfly house, this misty haven for native and exotic species draws visitors into an interactive experience. Beyond the 800 free-flying butterflies fluttering through the exhibit, there are educational stations where all ages can learn about the butterfly life cycle, migration patterns, how they find food and mates and elude predators. A large chrysalis case with several shelves holds native species waiting to emerge. This is one of the better emergence exhibits—it shows a wide variety of butterfly chrysalides in an easy to view display.

On any given day there are 20 top butterfly species that you will most likely see. Of the natives look for Black Swallowtails, Buckeyes, Cabbage Whites, Viceroys, Giant Swallowtails, Monarchs, Mourning Cloaks, Painted Ladies, Pearl Crescents and Red Admirals. The top ten exotics are Tiger Longwings, Blue & White Longwings, Tailed Jays, Small Postmen, Checkered Swallowtails, Blue Morphos, Red Crackers, Common Mormons, Clippers, and Great Egg Fly Butterflies. You can see photos of these butterflies and learn detailed information about their individual biology and habitats in the nature center's online **Butterfly Lab**. The lab has a complete list of the 40 North American species and 112 international species permitted to fly in the haven.

> The oldest known insect is a 350-million-year-old Russian fossil.

Look for two kinds of butterfly feeding stations: those with nectar and those with overripe fruit. Not all butterflies, such as the Owls and the Morphos, get their sustenance from flowers so the center provides the fruit. Before you leave look for the mirrors at the double door exits; they help you check for any hitchhiking butterflies, which are not allowed outside the house. If you still want to peek at the activity swirling inside, there are five porthole windows located on the haven's east wall, which allow you to see the habitat's interior from inside the museum.

Hours: 10 a.m. to 5 p.m. Sat.–Sun., 9 a.m. to 4:30 p.m. Mon.–Fri. Closed Thanksgiving, Christmas and New Year's Day.

Admission: $6 adults, $4 seniors and students with I.D., $3 children, free admission for all on Tue.

Children's Garden

Chicago Botanic Garden

1000 Lake Cook Rd., Glencoe, IL 60022
847-835-8208
www.chicagobotanic.org

Some **Big Bugs** are coming to this big botanical garden. They are part of New York artist David Rogers' monstrous exhibit of dinosaur-sized bug sculptures made of all natural materials such as trees, dried branches, roots, vines and bark. The **Big Bugs** exhibit has visited gardens across the U.S. including this one in 1998. This time the bugs are staying from June through late October 2002. You'll find them placed strategically throughout the garden's 385 acres: three 700-pound ants, two dragonflies, a ladybug, butterfly, grasshopper, beetle, earwig, Assassin Bug, a 1,200-pound Praying Mantis and an enormous spider poised for attack in a 15-foot web. Joining this gargantuan crew of critters is a new series of sculptures illustrating the important role insects play in pollination. They include a bee, beehive and flower.

Even without the **Big Bugs** there's plenty to be bugged about here. In the **Fruit and Vegetable Garden**—one of 23 separate gardens in this botanical oasis—you'll find an observation beehive plus a couple of regular working beehives. Accompanying the observation hive is an exhibit explaining the habits of bees and how they're important to gardens.

During the warmer months of May to September volunteers staff a **Discovery Station** from 11 a.m. to 3 p.m., Wed.–Mon. Look through a microscope to get a magnified view of a bee's wings, stinger and leg. Touch real beeswax and smell different types of honey. Learn about beneficial insects likes ladybugs, butterflies and other pollinators. Ask questions, play with hand puppets and maybe see some live bugs.

> Entomologists discover 12 new insect species daily.

Catch some butterflies over in the **Children's Garden** where butterfly-friendly plants are plentiful. Head for the

70

worm bins where kids can dig and find the tillers of the soil—earthworms. Kids can look at them but they have to put them back—they have a big job to do overturning earth so the plants can grow.

Hours: 8 a.m. to Sunset. Closed Christmas Day.

Admission: Free.

Parking: $7.75 per car, $5.75 for seniors on Tuesdays.

Hamill Family Play Zoo & Butterflies!

Brookfield Zoo

3300 Golf Rd., Brookfield, IL 60513
708-485-0263 or 800-201-0784
www.brookfieldzoo.org

This relatively new three-acre family zoo within the larger zoo is both an indoor and outdoor interactive adventure. The revolutionary new concept in zoo exhibits gets kids involved and caring about animals and nature. Bugs, of course, are an integral part of the zoo's habitats.

In the outdoor **Play Gardens** you'll find a 100-foot-long **Bug Walk** decorated with child-made insect art projects. There are live insects and other invertebrates located along the walkway. Look under logs, dig in the dirt or view an ant colony. There are huge bug sculptures to climb and buggy costumes to put on too.

Elsewhere in the **Brookfield Zoo** you'll find a bevy of bugs, mostly in the zoo's **Habitat Africa** and **The Swamp**: Brown Water Scorpion, Creeping Water Bugs, Salvage Beetles, Predatory Diving Beetles, Striped Knee Tarantula, Golden Silk Spider, Giant Centipede, Giant Prickly Sticks and Bess Bugs.

For 2002 the zoo is all aflutter with **Butterflies!** a seasonal habitat featuring more than 40 North American species. Landscaped with a variety of flowering annuals, perennials and grasses, the screened enclosure will be home to Buckeyes, Monarchs and several varieties of Longwings and Swallowtails. A separate admission charge applies.

Hours: 10 a.m. to 5 p.m. daily; extended summer and some weekend hours. (Butterflies!) May 25 to mid-September.

Admission: (General Zoo) $7 adults, $3.50 seniors (65+) and ages 3–11, free for ages 2 and under. (Children's Zoo) $1 adults, 50 cents seniors (65+) and ages 3–11, free for ages 2 and under. (Hamill Family Play Zoo) $3 adults, $2.50 member adults, $2 seniors (65+) and ages 2–11, $1.50 member seniors (65+) and members ages 3–11, free for ages 2 and under.

Underground Adventure

The Field Museum

1400 S. Lake Shore Dr., Chicago, IL 60605
312-922-9410
www.fieldmuseum.org

Visit the museum's unusual 15,000-square-foot **Underground Adventure** exhibit, where you walk through a soil ecosystem and encounter larger-than-life critters. It's a bit creepy when you feel you're the size of a bug and you meet animatronic creatures 100 times their normal size such as a four-foot-long centipede, a Mole Cricket clawing through a tunnel wall and a mother earwig protecting her brood. Lots of interactivities here, too: play with mud, manage a 1,000-acre farm on a computer and see clips from the **Underground Adventure Film and Video Festival**.

The museum has put together another buggy exhibit, **Insects: 105 Years of Collecting**, displaying a portion of one of the largest insect collections in the Western Hemisphere. See giant beetles and butterflies, 40-million-year-old insects in amber and more.

Hours: 9 a.m. to 5 p.m. daily, 8 a.m. to 5 p.m. Memorial Day to Labor Day. Closed Christmas and New Year's Day.

Admission: $8 adults, $4 seniors, students with I.D. and ages 3–11. Free on Mondays and Tuesdays, September to February.

Indiana

Butterfly Jungle Indonesia Rain Forest

Fort Wayne Children's Zoo

3411 Sherman Blvd., Fort Wayne, IN 46808
260-427-6800
www.kidszoo.com

Tropical butterflies are flying at one of the top kids' zoos in the U.S. Visit the **Indonesia Rain Forest** where orangutans, Sumatran Tigers and rare Komodo Dragons play.

Owl, Great and Common Mormon, Zuleika, Cyrgia, Common Rose and Zebra Longwing species can be found near the large windows of the **Butterfly Jungle**. Watch for eggs on the host plants. Look fast because the zoo staff carefully removes them so the caterpillars won't hatch and eat all the exhibit plants.

Above the waterfall look for the **Butterfly Hatching Area** where new winged wonders wait in their chrysalides pinned to the shelves.

Hours: 9 a.m. to 5 p.m. daily, late April through mid-October.

Admission: $6.50 adults, $4 seniors (60+) and ages 2–14, free for ages under 2.

> Insects account for 85% of all known animal species.

Iowa

Insect Zoo

Iowa State University

407 Science II, Ames, IA 50011
515-294-4537
www.ent.iastate.edu

Room 407 in **Science II** at **Iowa State University** doesn't look like much, but what's inside will make your eyes bug out. Established as an onsite **Insect Zoo**, the room brims with more than 50 species of bugs. Run by the university's world-renowned Entomology Department, the **Insect Zoo** is a feast for bug lovers. Among the main attractions: Tropical Millipedes, New Guinea Walking Sticks, West African Assassin Bugs, Malaysian Centipedes, Goliath Bird Eater Tarantulas, Cave Roaches and Madagascar Hissing Cockroaches.

Native mainstays include Iowa Walking Sticks, Chinese Mantids, Super Worms, ladybugs, Bess Bugs and German and American Roaches. Look for these aquatic insects: Water Scorpions, Giant Water Bugs and Whirligigs. In the non-insect collection, the zoo shows off Carolina Millipedes, Pumpkin Millipedes, Pink Toe Tarantulas, Curly Hair Tarantulas, Mexican Red Leg Tarantulas, Wolf Spiders and Orb Weaver Spiders. The number of invertebrates balloons to about 100 species in the spring and summer, when entomologists and students go on collecting trips.

The only hitch? You have to make a reservation two or more weeks in advance. The benefit? You get an incredible, individualized hands-on session with the bugs. As part of your zoo visit you'll get to hold a large number of the insects and arthropods in the exhibits. A trained staffer assists and guides you during your visit.

Tours and the hands-on activities that accompany them are designed to fit your group's needs whether you're a family of two or a group of 60 elementary students. For educational tours, hands-on activities can include simple dissection of insects, bug safaris in the campus prairie areas and collection workshops.

Better yet, if you arrange to come in the spring or summer, the zoo will arrange to have fresh, live butterflies flying in it's seasonal butterfly house. The house, located across the street from the zoo, is set up only when a tour is expected. The **Insect Zoo** releases about 20 to 30 butterflies into the enclosure. Among the species you might see: Zebra Longwings, Julias, Monarchs, Regal Fritillaries, Red Spotted Purples, Luna Moths and Tiger Moths.

And if you're in the Ames area during September, let the university entomology club entertain you at their annual insect fair (see Bugged Out chapter).

Kansas

Kansas State University Butterfly Conservatory & Insect Zoo

Kansas State University Garden

Denison Ave.
785-532-6154
www.ksu.edu/butterfly
insect@ksu.edu

Bugs are swarming **Kansas State University** and it's a showcase of hoppers, creepers and flyers you won't want to miss. After roaming without a true home for years, the **Entomology Department**'s **Insect Zoo** is re-emerging in new, state-of-the-art digs at the renovated **Quinlan Visitor Center**, the former caretaker's cottage at the **Old Dairy Barn**. The 1,400-square-foot room adjacent to the rose garden is simply buzzing with activity under the direction of entomology department chair Dr. Sonny Ramaswamy and professor Dr. Ralph Charlton.

The two have led the way for the zoo to become a wonderful place for families to discover the delight of live bugs, from centipedes to millipedes to scorpions. Inside you'll find some innovative bug displays such as the tropical tree trunk with cutouts that contain a variety of live insects. Look closely at the plastic tubes winding around the room and you'll see Leaf Cutter Ants crawling through, carrying the bits of leaves they need to cultivate their fungus farm. There's even a walk-in cave with motion detectors that light up glow-in-the-dark insects such as scorpions.

See how bees build comb, make honey and lay eggs by peering into an observation Honey Bee hive. A water exhibit features Poison Dart Frogs, fish and aquatic insects such as Diving Beetles and Giant Water Bugs. The **Bug Doctor** "is in" for a couple of hours each day and his (or her) door is open for free consultations. Bring your big bug problems or bring in a bug you want identified. You can also get bug help by emailing bugs@ksu.edu. Tarantulas are a favorite here since Dr. Charlton is an expert, so you'll see more than a dozen of these amazing arachnids. Rounding out the zoo are several species of stick insects, leaf insects and cockroaches. A Web cam will be up in the future so check the zoo's online site periodically.

From here it's not a big leap to the gardens' glass conservatory, a Victorian structure originally called the **Plant Museum** when it was built in 1907. Through a cooperative effort with the university's Horticulture Department, butterflies were brought to life at the university gardens. They were temporarily derailed a while ago due to a bad hailstorm, but the butterflies are definitively flying again. Watch as about 300 exotic and native butterflies such as Julias, Malachites and Monarchs flutter among the Passiflora, Banana Trees and Bougainvillea in the tropical rain forest in the building's west wing. Toward the end of the day find the Monkey Puzzle tree where Zebra Longwings like to cluster. A significant butterfly to look for is the Albino (White) Monarch supplied by **KSU** professor Orley "Chip" Taylor who runs **Monarch Watch**, an international outreach group for these migrating marvels.

Outside the conservatory visit the stunning gardens set up in themes such as herb, iris and yes, a butterfly one. See if you can find the bronze statue of a young man sitting reading a book with a butterfly resting on the tip of his finger.

Limited parking is available so call the KSU Parking Service at 785-532-7275.

Hours: (Insect Zoo) 10 a.m. to 4 p.m. daily. (Butterfly Exhibit) 10 a.m. to 4 p.m. Mon.–Fri., 11 a.m. to 2 p.m. Sat.–Sun. February to November.

Admission: Free.

Butterfly House

Botanica, The Wichita Gardens

701 Amidon, Wichita, KS 67203
316-264-0448
www.botanica.org

Every June fluttering rainbows open their wings and fly free in this 2,800-square-foot Great Plains butterfly house. Hundreds of regional butterflies weave their way through the net-covered enclosure's botanical bouquet. Sulphurs, Swallowtails, Monarchs, Queens and Zebra Longwings are among the 200 to 300 chrysalides that arrive each week waiting for their chance to take flight. More than 5,000 are released throughout the summer and until the first fall frost. You can take a peek at their emergence in the special butterfly-hatching house.

See more native and migratory butterflies zip through **Botanica's** outdoor **Butterfly Garden**, planted with flowering nectar plants and food plants. Look for the puddling fountain that provides water for the butterflies. Small sitting areas along the garden's brick path provide nice places to pause and observe nature.

Located in the **Museums on the River District**, there's plenty more to explore in this botanical bounty. Theme gardens include a **Wildflower Meadow** and **Wildflower Woodland**, a **Shakespearean Garden**, a **Rockery Garden** and an **Aquatic Garden**.

Hours: 9 a.m. to 5 p.m. Mon.–Sat., 1 p.m. to 5 p.m. Sun. and holidays, mid-March through December; until 8 p.m. Tuesdays June through September; 9 a.m. to 5 p.m. Mon.–Fri. January to mid-March. Closed Thanksgiving Day, Christmas Day and New Year's Day.

Admission: $6 adults, $5 seniors (62+), $3 children, free for ages 5 and under, $12 family rate for parents and all minor children under age 21.

Kentucky

Arachnamania!

Louisville Zoo

1100 Trevilian Way, Louisville, KY 40213
502-459-2181
www.louisvillezoo.org

Love hairy eight-legged critters? Then meet the world's largest spider, the Goliath Bird Eater from South America and its cousin the Purple Bloom Bird Eater. These and at least nine other spider species call the zoo their home. **Arachnamania!** is truly for the arachnamaniac, with an emphasis on Theraphosid spiders (tarantulas). The exhibit can make some people feel creepy, although tarantulas in particular are actually shy and most are not poisonous to people. They'd prefer to hide under a rock rather than use their large fangs in self-defense. Most are solitary creatures since females will try to eat their mates. The exceptions are the Peruvian Pink Toe Tarantulas, which are housed in a community setting at the zoo. Other eight-leggers include the King Red Baboon Spider from Kenya, the Horned Baboon Spider from South Africa, the Costa Rican

Zebra Spider with zebra type stripes on its legs and the Western Blonde Tarantula, which crosses Southwestern highways in search of females.

Hours: 10 a.m. to 4 p.m. September to March, 10 a.m. to 5 p.m. April to August.

Admission: $8.95 adults, $6.95 seniors (60+), $5.95 ages 3–11, free for ages 2 and under.

Louisiana

Audubon Insectarium (opening 2003)

Audubon Nature Institute

#1 Canal St., New Orleans, LA 70130
504-861-2537
www.auduboninstitute.org

The nation's only freestanding insect museum will open in Spring 2003 if plans to construct the 30,000-square-foot facility continue to move forward. Operated by the **Audubon Institute** which also runs the **Aquarium of the Americas**, and the **Audubon Louisiana Nature Center**, the insectarium will rival almost any bug exhibit in the U.S.

Located on historic Canal Street near the famous French Quarter, the insectarium will house both live and preserved specimens with interactives that kids will go buggy for. Among the exhibits:

A **Collector's Camp** with a bug identification station and lab staffed by an entomologist.

A kitchen where insect recipes can be sampled.

An insect aviary with free-flying bugs.

A multimedia experience with videos.

Interactive stations about New Orleans insects, plant and insect relationships, pre-historic bugs and unusual insect behavior.

Hours and admission to be determined. In the meantime, check out the **Incredible Edible Insect Event** at the **Audubon Louisiana Nature Center** in October (see Bugged Out chapter).

Maine

Explore Floor

Children's Museum of Maine

142 Free St., Portland, ME 04101
207-828-1234
www.kitetails.com

Transport the kids to other worlds at the only place in Maine known to harbor live bugs. On the second floor, thousands of Leaf Cutter Ants populate glass pods exposing the life of these insects that farm fungus for food. The ants carry bits of leaves on their backs then turn them into mulch and grow the fungus. What's particularly cool about the exhibit is that you can crawl underneath it and stick your head inside a glass cube and view the ants up close.

Sometime in early 2002, the museum plans to open up a new permanent exhibit on camping, which will also feature some displays on bugs.

Don't miss the **Camera Obscura**, one of only three in the U.S. You'll get a panoramic view of the city and the waterfront.

Hours: 10 a.m. to 5 p.m. Tue.–Sat., Noon to 5 p.m. Sun., members-only Toddler Time 9 a.m. to 11 a.m. Mon., Labor Day to Memorial Day; 10 a.m. to 5 p.m. Mon.–Sat., Noon to 5 p.m. Sun., Memorial Day to Labor Day. Open until 8 p.m. first Friday of every month.

Admission: $5 per person, free for ages under 1 year. Free first Friday night 5 p.m. to 8 p.m. every month.

Maryland

Wings of Fancy

Brookside Gardens

1800 Glenallan Ave., Wheaton, MD 20902
301-949-8230
www.brooksidegardens.org

Fanciful wings of color take flight in the **South Conservatory** during the annual butterfly event at this 50-acre display garden. Surround yourself with 20 to 25 species of North American butterflies that navigate their way among butterfly-attracting annuals, tropical varieties and hardy shrubs. A net is dropped inside the 62' by 42' space so that butterflies stay low.

Be dazzled by the blue hues of native Maryland butterflies such as Red Spotted Purples, Pipevine Swallowtails and Eastern Black Swallowtails and the orange colors of Red Admirals, Question Marks, Viceroys, Monarchs and the Baltimore, the official state butterfly. Other locals include Painted Ladies, Buckeyes, Zebra Swallowtails, Hackberries, Palamedes Swallowtails and Great Purple Hairstreaks.

Learn how to become a butterfly gardener by planting flora native to the mid-Atlantic region. Discover the marvel of metamorphosis by getting a close eye view of butterfly eggs on host plants inside the conservatory. Docents help you locate both eggs and caterpillars.

Hours: 10 a.m. to 4 p.m. daily, mid-May to late September.

Admission: (Butterfly show) $3 per person (2+). (Gardens) Free.

> There are 165,000 species of butterflies and moths.

Massachusetts

The Butterfly Place

120 Tyngsboro Rd., Westford, MA 01886
978-392-0955
www.butterflyplace-ma.com

Come inside this 3,100-square-foot glass atrium where locals mingle with exotics. We're talking butterflies, about 50 varieties, from native Spicebush Swallowtails to exotic Blue Morphos. At any one time 500 rainbow wings may flutter around you or up to the 27-foot peak of this living butterfly environment.

The butterfly house likes to feature as many local varieties as possible such as Baltimore Checkerspots and Great Spangled Fritillaries seen mostly in June and July, as well as Zebra Swallowtails, Mourning Cloaks, Question Marks, Commas and Cabbage Whites. Wind through the nectar plants on a set pathway and see popular tropicals such as Owls, Julias, Red Lacewings and dozens of Swallowtails and Longwings. Wear some body lotion and the atrium's black and white Paper Kite Butterfly, also known as the Idea from the Philippines, won't leave you alone.

Got questions? Then head for the **Show & Tell Bench** in the middle of the atrium where a staff member can provide answers. Meet some other guests there such as Praying Mantids, walking sticks and Madagascar Hissing Cockroaches. In July through September, the bench usually features the seven-inch long, spiky Hickory Horndevil caterpillar, the precursor to a Royal Walnut Moth.

You can also view the kaleidoscope of colors through the windows of an observation room where several cases of live butterflies and moths in different stages of development are on display. A 15-minute video about a butterfly's life cycle also runs continuously in this area.

If you're ever in Branson, Missouri, be sure to visit the big sister of this facility, a 9,000-square-foot greenhouse featuring tropicals and Southern butterflies.

Hours: 10 a.m. to 4 p.m. March 1 to March 31; 10 a.m. to 5 p.m. April 1 until Columbus Day. Closed Easter Sunday.

Admission: $7.50 adults, $6 seniors (65+), $5.50 ages 3–12, free for ages 2 and under.

Magic Wings Butterfly Conservatory & Gardens

281 Greenfield Rd., South Deerfield, MA 01373
413-665-2805
www.magicwings.net

Wing your way to one of the largest year-round butterfly houses in the Northeastern U.S. As you step in from the frosty cold into the 4,000-square-foot glass **Frances R. Redmond Conservatory** you're instantly warmed by its 80-degree tropical environment and the beauty of bountiful butterflies. Between 500 and 1,000 tropical and native species soar to the sounds of water and peaceful music and flutter among the Paw Paw, Sassafras and Asters.

More than 60 varieties of flying flowers are hatched, imported or bought from local butterfly farmers. There are Gossamer Wings such as Bog Coppers, Eastern Tailed Blues and Spring Azures; Anise, Pipevine and Gold Rim Swallowtails; Brushfooted Butterflies such as Dianas, Ruddy Daggerwings, Malachites, Blue Wings and Red Admirals; Silver Spotted, Long-Tailed and Cobweb Skippers; Sulphurs such as Eastern Dogfaces, Sleepy Oranges and Cabbage Whites; Milkweed Butterflies such as Monarchs, Queens and Soldiers; plus Common Wood Nymphs and Tawny Emperors. You'll find a complete list on the conservatory's Web site.

Sit on the carefully placed benches and watch Tiger Longwings and Tiger Swallowtails spiral around the heart-shaped Koi pond. "Flight

Attendants" stationed throughout the conservatory can help you identify the Buckeyes and the Baltimores. They'll also show you where the native butterflies have laid eggs on the host plants.

The warmth of the conservatory actually begins before you ever enter the glass house. It's part of a 14,400-square-foot facility that includes a plant-filled atrium and a garden railroad, the **Butterfly Express**, which is particularly fun for birthdays. Warm yourself by the fieldstone fireplace on wintry days and admire some of the butterfly artwork hanging in the admissions area. Look for the 12' butterfly kite, a quilt by dottie case, a poster of the **Butterfly Alphabet** by Kjell B. Sandved and **Butterflies USA**, an original painting by Alan James Robinson. There's more to admire in the conservatory's **Camberwell Art Gallery** and the gift shop.

Visit the outdoor **Iron Butterfly Garden** in the warmer months for a field full of dancing native butterflies. There are gazebos, quiet nooks and even a garden maze to walk through. Make it a complete day with lunch at the **Flying Rainbows Café**.

Hours: (Winter) 9 a.m. to 5 p.m. (Summer) 9 a.m. to 6 p.m.

Admission: $7 adults, $5 seniors (62+), $4.50 ages 3–17 (up to age 22 with student ID), free for ages under 3.

Butterfly Landing

Franklin Park Zoo

1 Franklin Park Rd., Boston, MA 02121
617-541-LION
www.zoonewengland.com

Walk into one of the largest seasonal live butterfly exhibits in North America and you're treated to clouds of about 1,000 butterflies in free flight. The tented outdoor exhibit measures 174' long by 42' wide by 21' high. Look for up to 45 North American species gently swaying to soothing Surround Sound.

Dancing among the Bee Balm, Marigolds and Zinnias (nectar plants) are Massachusetts' natives such as Mourning Cloaks, Monarchs, Question Marks, Red Spotted Purples, Cloudless Sulphurs, Sleepy Oranges, Gulf Fritillaries, Viceroys and Painted Ladies. Several species of swallowtails soar through the exhibit: Eastern Tigers, Pipevines, Giants, Spicebushes, and Eastern Blacks. Look for their cousins, the exotic Palamedes and Zebra Swallowtails as well as Julias, White Peacocks and Zebra Longwings.

A colorful zoo pamphlet helps you identify some of the species and corresponds with field guides located through the exhibit. Look for the emerging box to view chrysalides in several stages of growth. About 700 butterfly pupae arrive weekly from different butterfly farms in Texas, Tennessee and Florida.

Since this seasonal exhibit is exposed to the elements, it may not be open when you visit the 72-acre zoo located in an historic Boston park. Check the "butterfly meter" in front of the exhibit to see if butterflies are in flight that day.

Butterfly Landing opens Memorial Weekend so look for festivities and activities surrounding the event.

Hours: 10 a.m. to 5 p.m. April 1 to September 30 (until 6 p.m. weekends & holidays), 10 a.m. to 4 p.m. October 1 to March 31.

Admission: (Zoo) $9.50 adults, $8 seniors, $5 ages 2–15, free for ages under 2. (Butterfly Landing) $1 per person.

Michigan

Foremost's Butterflies Are Blooming

Frederik Meijer Gardens

1000 E. Beltline NE, Grand Rapids, MI 49525
616-957-1580 or 888-957-1580
www.meijergardens.org

Spring is in the air and the butterflies are blooming at the largest temporary butterfly exhibit in the U.S. For two months more than 35 species of dazzling fluttering flowers from South and Central America wing their way through the 15,000-square-foot tropical environment at the **Lena Meijer Conservatory**. Look for harvest-moon-orange Julias, tiger-striped Ismenius Longwings, jewel-toned Malachites and electric Blue Morphos. See if you can spot the Blue Wings that resemble stained glass, the Tropical Swallowtails that look like they have white eyes and a red mouth and the Postman Butterflies that sport a white mustache when they flatten their wings.

In 2002 the exhibit celebrates its seventh year at the 125-acre botanical gardens and sculpture park. More than 10 species of Asian butterflies native to Malaysia, Singapore, China and Burma have been added to the exhibit including Scarlet Mormons, Lime Swallowtails and Emerald Swallowtails. Known for their large size and expansive wings they bring big brilliant bursts of color.

Visit the **Butterfly Bungalow** where new chrysalides are added weekly and where the miracle of metamorphosis unfolds every day. Watch too, as the adults feed at nectar stations throughout the conservatory. Purchase a colorful butterfly identification booklet as a souvenir and learn more about butterflies and their habitats. Look for the butterfly Web cam to see who's flapping their wings before the camera.

Head for the **Family Fun Center** where kids can participate in a hands-on game and experience a larger-than-life display of the butterfly life cycle. On weekends from Noon to 3 p.m., children can create "make-and-take" projects. Look for the opportunity for kids to start a new card collection with full-color butterfly cards that feature stats and fun facts about specific species. Each child age 13 and under receives one free card with paid admission. Also look for the **Michigan Butterfly Garden** in the **Grace Jarecki Seasonal Display Greenhouse**.

Be sure to go online and see the Web cam in action. While the camera is only operative during the exhibit, the gardens and Foremost maintain a yearlong educational site on butterflies. There are lots of learning activities including how to create a backyard butterfly garden, butterfly postcards to email and instructions for creating your own flutterby.

Hours: 9 a.m. to 5 p.m. during butterfly exhibit.

Admission: $8 adults, $7 seniors (65+), $6 students with I.D., $4 ages 5–13, free for ages 4 and under. Includes admission to the entire gardens.

What's the Buzz?

Grand Rapids Children's Museum

22 Sheldon Ave. NE, Grand Rapids, MI 49503
616-235-4726
www.grcm.org

Make a beeline to this cheerful hands-on children's museum to watch live bees buzzing in an observation hive. The clear Plexiglas® home is located on the second floor and vents to the outside so the bees can buzz to their pollen and bring it back. Watch as the bees build their honeycombs, manufacture wax, store food and tend to their nursery. Spot the white dot on the queen using a magnifying glass.

Wannabee a bee? Kids can dress up in bee costumes, play in an interactive hive and view magnified hive photo books. They can also try on a beekeeper's outfit, play with bee puppets, taste locally produced honey and handle a big chunk of sweet-smelling raw beeswax.

Hours: 9:30 a.m. to 5 p.m. Tue., Wed., Fri., Sat., 9:30. a.m. to 8 p.m. Thu. (5 p.m. to 8 p.m. Family Night, $1 admission), Noon. to 5 p.m. Sun.

Admission: $3 per person.

Bug House

Michigan State University

147 Natural Science Building, East Lansing, MI 48824
517-355-4662
www.ent.msu.edu

Welcome to the **MSU Department of Entomology's Bug House**. Located in rooms 146-147 of the **Natural Science Building**, the **Bug House** is cozy and entomology students who know their bug stuff give you a warm welcome. These buggy volunteers guide you to a collection of live and pinned insect displays. The preserved specimens have been collected at the university since just after the Civil War and now their numbers have exceeded one million. For live bugs you'll be treated to a buzzing hive full of busy bees. Look at the different frames of the hive through its Plexiglas enclosure and watch as the bees work their way in and out of the tunnel that leads outdoors. The **Live Room** has a standard set of bugs you can handle such as walking sticks, hissing cockroaches and millipedes. Tarantulas are here too, but they're too frisky to fish out. Arrange for a one-hour tour by calling ahead or visit the house of bugs during the free public hours Monday evenings.

Look for the campus **Butterfly House** at the **Plant and Soil Science Building** across from the **4-H Children's Garden**. Call 517-355-0348 to arrange a tour.

Hours: (Bug House) 9 a.m. to 3 p.m. Mon.–Fri. for pre-arranged tours; 5:30 p.m. to 7:30 p.m. Mondays for public tours. Closed in December. (Butterfly House) 9 a.m. to 5 p.m. Mon.–Fri. or by appointment.

Admission: (Bug House) $30 for guided tours of 30 or fewer participants. Free on Monday nights. (Butterfly House) $2 adults, $1 children, $20 minimum per group.

Wildlife Interpretive Gallery

Detroit Zoo

8540 West Ten Mile Rd., Royal Oak, MI 48068
248-398-0900
www.detroitzoo.org

Come early and beat the crowds to the zoo's oldest building, the **Wildlife Interpretive Gallery**, built circa 1928 and located near the front entrance. Access is limited so you may have to wait in line to enter the gallery's live **Butterfly/Hummingbird Garden**. Walk along the circular pathway and about 30 different species of farm-raised butterflies, from Central America and Malaysia, flutter around you. As you walk underneath the glass-domed building you'll see 300 to 400 of these tropical beauties such as iridescent Blue Morphos and Tiger Swallowtails. The former bird house has benches, a fountain created by sculptor David Barr and is a popular place for weddings—the zoo director had his own here. The gallery includes a wildlife art display area, a theater, a coral reef aquarium and interactive display. Look for the nautilus shell pattern on the rotunda floor. From here you can walk to the **Free Flight Aviary** with 30 species of tropical birds.

Hours: 10 a.m. to 4 p.m. daily, November to March.; 10 a.m. to 5 p.m. daily, April to October; 10 a.m. to 6 p.m. daily, Sundays, holidays mid-May to Labor Day; 10 a.m. to 8 p.m. Wednesdays, late June to August.

Admission: $8 adults, $6 seniors (62+) and ages 2–18, free for ages under 2.

World of Spiders

Belle Isle Zoo

Belle Isle Park, E. Jefferson at E. Grand Blvd., Detroit, MI 48207
248-398-0900
www.detroitzoo.org

The world's largest exhibit of live spiders is floating in the middle of the Detroit River. Perched on the 1,000-acre Belle Isle Park, is a 13-acre zoo whose most requested educational exhibit are the spiders. There are more than 30 species of spiders here so you'll see more than your typical Mexican Red Leg Tarantula, although that's here too. How about a Costa Rican Zebra Spider with Zebra-like stripes on its legs? Or the Mombassa Golden Starburst, which ordinarily makes a home under the raised floors of houses in East Africa? Or the Bluebloom whose hair has a slight bluish cast? Want more? See the Belize Cinnamon whose velvety outside looks like the spice and the aggressive Amazon Orange Banded Tarantula.

The 20' by 30' exhibit is located just inside the zoo's entrance on the right-hand side and includes spiders from all the world's regions except for Antarctica (where no spiders apparently exist). Each spider lives in a miniature version of its own native habitat so you get to see what life is like for these arachnids outside the zoo setting. A new display shows the molting process that spiders go through in order to grow. Next to the exhibit you'll find some other small invertebrates like millipedes and centipedes.

Hours: 10 a.m. to 5 p.m. daily, May 1 to November 1 only.
Admission: $3 adults, $2 ages 2–18, free for ages under 2.

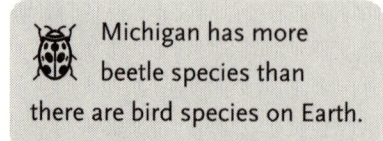

> Michigan has more beetle species than there are bird species on Earth.

The Bug House

Kalamazoo Nature Center

7000 N. Westnedge Ave., Kalamazoo, MI 49009
616-381-1574
www.naturecenter.org

If the 1,000 acres of rolling hills with its 11 nature trails isn't enough to draw you in, the conservatory of live buggies will. **The Bug House** is an old converted greenhouse and its crawling residents live in terrariums that stand on raised plant beds. You'll find between 8 and 15 species of arthropods here, depending on the time of year. Most likely you'll see Vietnamese Centipedes, Vinegaroons, Black Emperor Scorpions, Giant African Millipedes, Darkling Beetles, Rosy Tarantulas, Praying Mantids, ladybugs, aphids, assorted beetles and Hermit Crabs. A 120-gallon aquatic tanks holds dragonfly larva, Giant Water Bugs, Ferocious Water Bugs, Sunburst Diving Beetles, Spotted Diving Beetles and Water Scorpions that look like a cross between a scorpion and a walking stick. Snails, bullfrogs and tadpoles also inhabit the tank.

The center's 11-acre **Arboretum** has a **Hummingbird-Butterfly Garden** that's in bloom from June to October. There's also a 25' by 12' butterfly house where up to 10 native species fly free. The house is stocked with 50 new chrysalises every two weeks so you should see between 50 and 100 flying beauties: Tiger Swallowtails, Zebra Longwings, Gulf Fritillaries, Viceroys, Queens, Monarchs, and Julias. Old potting beds have tropical plants, a pond is filled with Koi and Pacu fish, and you can find frogs in the bubbling fountain. A neat aspect—Praying Mantids and ladybugs roam free as pest controllers.

Hours: (Nature Center) 9 a.m. to 5 p.m. Mon.–Sat., 1 p.m. to 5 p.m. Sun. (Bug House) Late May to November or when the cold weather starts.

Admission: $4.50 adults, $3.50 seniors (55+), $2.50 ages 4–13, free for ages 3 and under.

Minibeast Zooseum and Education Center

6907 West Grand River Ave., Lansing, MI 48906
517-886-0630
http://members.aol.com/YESbugs/zooseum.html

Visit the largest live insect and spider museum in Michigan. The 3,000-square-foot facility was started a few years ago by Gary and Dianna Dunn, founders of the **Young Entomologists Society** (YES). The combination zoo/natural history museum/nature center is a tribute to minibeasts. The Dunns have upwards of 40 arthropod species in their **Zooseum**, which is actually located on the lower level on their home on a seven-acre lot. About a dozen or so of the creeping critters are out on display in 15-gallon aquaria tanks. When you walk in you may find Malaysian Centipedes, African, South American and Texas Millipedes, Apple and Winkle Snails, Red Worms, Black Widows, Wolf Spiders, Velvet Ants, Tiger Beetles, Whip Tail Scorpions, Death's Head Cockroaches, Death Feigning Ladybird Beetles, and any number of tarantulas. Popular with kids are the Sow Bug and Eastern Subterranean Termite colonies.

Kids will have fun playing with the interactive exhibits like **Bug Tic Tac Toe** and the **Insect Roller Course**, a combination of skeetball and pinball. There are also computers where kids can play insect games and use bug-related software. Listen for the musical butterfly clock that flaps its wings every hour on the hour.

Outside is a mile-long trail where you can take a bug safari through various habitats. Interpretive signs mark the major tree species and flowering plants. Look for the Dunns' insect collecting devices along the **Buttonbush Swamp Trail**.

If you're a member of **YES** or the **Zooseum** you can use the **Resource Center and Library** which houses a large collection of insect related books, posters, video, computer programs, photos, artifacts and toys. It's easy to join **YES**. Just visit their Web site (see Be a Bugatist chapter).

Hours: 1 p.m. to 5 p.m. Tue.–Fri., 10 a.m. to 5 p.m. Sat., early March through December. Closed Sunday, Monday, Thanksgiving and Christmas. Closed December 26 to early March.

Admission: $3.50 adults, $2.50 seniors (60+) and ages 3–12, free for ages under age 3.

Zooseum Annual Pass and YES membership: $25 individual, $55 family.

Mackinac Island Butterfly House

P.O. Box 1308, Mackinac Island, MI 49757
906-847-3972
www.mackinac.com/butterflyhouse

Postcards don't get any prettier than this Great Lakes isle and its butterfly house. A mere three miles by two miles in size, **Mackinac Island** (pronounced *Mackinaw*) is lush, historic and off-limits to cars. The 1800-square-foot **Butterfly House** is located near the **Grand Hotel** and flies up to 1,000 winged wonders at any one time. One of the first butterfly houses built in North America, the glass-enclosed greenhouse has permits to display about 450 different species; about 70 species can be seen in any one season. Among the best of the beauties are the vibrant Blue Morphos and the black and green Triodes Brookianas, endangered butterflies also known as Rajah Brooke's Birdwings. The size of the house coupled with the number of flying butterflies means a chance to really be surrounded by a fanning kaleidoscope of color. There's also an emergence area to watch the wet wings come out of their chrysalides.

Hours: Daily, mid-May to early October

Admission: $5 adults, $2 ages 6–12, free for ages 5 and under. Paid admission covers as many visits as you like while you stay on the island.

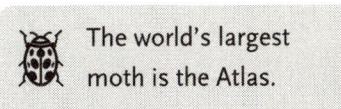

> The world's largest moth is the Atlas.

Minnesota

Science Museum of Minnesota

120 W. Kellogg Blvd., St. Paul, MN 55102
651-221-9444
www.smm.org

And now for something completely different ... Dermestids, tiny beetles that eat the flesh of the less fortunate. You can see them in action through a special large viewing window built into the wall of the museum's **Biology Lab** on Level 3. So when you stand in line for the **Laser Theater**, you'll get an extra show for your ticket.

It sounds gruesome, but there's only dried flesh on the bones that these black bugs and their larva are gnawing on. The beetles are actually a biological tool that hundreds of other museums use, only visitors to this science center get a rare look at these entomological eaters at work. The bones are from specimens that scientists preserve by stripping off the skins (later used as study skins). When the bones are clean, they are kept in the museum's osteology or bone collection.

Dermestids like it hot and humid so they have their own environmental system in the **Biology Lab**. And their room is well sealed because if they were to escape, they would devour the museum's display collections. For pest control, the biologists put the bones in the deep freeze before they leave the lab. On a related note, the museum celebrates **National Pest Control Month** in June, so look for a variety of activities about bugs at that time.

You can watch and talk to scientists as they do their work, including the skinning of animal skeletons, at the museum's **Visible Lab**. It's located in the **Collections Gallery** on Level 4 where you can see a video on Dermestids. Be sure to look for the display of 165 native Minnesota butterfly specimens. They're arranged according to family groups and have information on the date and location of the collection. The gallery is devoted to describing why scientists collect, how they collect and how they preserve things like butterfly specimens. It's an important subject for a science museum that has a collection of 1.75 million artifacts and hundreds of hungry beetles.

Hours: 9:30 a.m. to 5:30 p.m. Mon.–Wed.; 9:30 a.m. to 9 p.m. Thu.–Sat.; 10:30 a.m. to 5:30 p.m. Sun. Closed for one week during mid-September for annual cleanup.

Admission: $7 adults, $5 seniors (65+) and ages 5–12 for exhibits only.

Zoolab & Butterfly Garden

Minnesota Zoo

13000 Zoo Blvd., Apple Valley, MN 55124
952-431-9500
www.mnzoo.org

When you visit this 500-acre zoo in the summer, the Zebra you might see may not be a mammal. Instead it could be a Swallowtail Butterfly winging its way around the zoo's seasonal butterfly garden, located just west of the zoo's main entrance. The 80' by 30' netted garden is a huge hit and attracts hundreds of visitors per day who come to see the 16 butterfly species and 5 moth species that are released during the season. Not all species fly at the same time, so visitors return to see what's new. Look for Milbert's Tortoise Shells, Red Admirals, Red Spotted Purples, Pearl Crescents, Purplish Coppers and Sleepy Oranges. About 400 to 500 butterflies can be seen during the peak time. Watch them while sitting on one of several benches.

A **Chrysalis Tree** and a **Cocoon Log** give you a close look at the birth of butterflies and moths as they emerge from their pupae. There are about 75 to 100 chrysalides that hang from the tree at any one time and 50 to 100 cocoons on the log. If you've got a question about which species is emerging, trained staff and volunteers are available to answer questions.

Be sure to look for a Monarch caterpillar display during mid-summer which focuses on the life cycle of these majestic migrators. The Monarch is particularly important to Minnesotans because it was named the state butterfly in the year 2000.

Just around the corner from the zoo's main entrance is a hands-on discovery place called **Zoolab** where visitors can touch and interact with a wide variety of animals, including invertebrates. The lab is home to breeding colonies of Madagascar Hissing Cockroaches and African Giant Millipedes. The zoo has been very successful with the millipedes, raising one called Debris, that was so large at 13 inches long that it made the *Guinness Book of World Records*. Debris is no longer alive, but that's typical of a species that reaches its senior year at age 2. You can also see Orange Knee and Chilean Tarantulas here, but you have to make a reservation.

Hours: 9 a.m. to 6 p.m. daily, Memorial Weekend through Labor Day; 9 a.m. to 4 p.m. weekdays, 9 a.m. to 6 p.m. September; 9 a.m. to 4 p.m. daily October 1 to April 30; 9 a.m. to 4 p.m. weekdays, 9 a.m. to 6 p.m. weekends May. (Butterfly Garden) Memorial Day through Labor Day.

Admission: $10 adults, $6.25 seniors (65+), $5 ages 3–12, free for ages 2 and younger.

Missouri

Monsanto Insectarium

Saint Louis Zoo

One Government Dr., Saint Louis, MO 63110
314-781-0900
www.stlzoo.org

Learning from its predecessors, the **St. Louis Zoo** has created the Goliath (Beetle) of live insect exhibits in the U.S. With more than 100 species of invertebrates, 20 exhibit areas, a butterfly wing and a Missouri meadow, this place will bug your socks off.

Open only since 2000, the zoo's insectarium is always abuzz with its bug bounty. First you're greeted by an eight-foot-long realistic sculpture of a Centaurus Beetle that beckons you to a path through the bug building. In the center is a life-size simulated tree that reaches through the ceiling so that the bees from the hive inside can buzz to and from the outside. The honeycomb is visible through a window cut in the tree.

Learn how bug keepers and entomologists work by talking to them in person. Watch them at work through a window in the research area and ask questions via a two-way microphone.

There are so many buggy things to do that it's hard to decide which to do first. Here's a honey sampler of what the zoo calls its "Oh, Wow!" exhibits:

Not Home Alone: Lift the lids and open the drawers in the kitchen, front porch, yard and garden to see what bugs live with you at home.

Dune Buggies: Touch the rock to experience desert heat then bury your hand in a burrow to experience the cool desert night. Discover how some insects live in both extremes.

Winging It: Spin the wheel to see how different insects, such as dragonflies and butterflies, wing their way through the air.

Buzz, Blink, Chirp, Hiss, Sniff: Find a date in the dark as you try to match a female firefly's flash pattern with that of her mate. Discover how other bugs communicate.

After all this, what live bugs will you see? Plenty. Here's a look at a few of the insectarium's most intriguing invertebrates:

Honey Pot Ants: A Southwestern desert resident, these ants store food as a sugary liquid then bring up the nectar to feed others when food is scarce.

Orchid Mantis: This tropical insect can hide in plain sight if it sits on a rain forest orchid.

Prairie Mole Cricket: It builds a "stereo system" of tunnels then broadcasts its call for a mate.

Fleas: Not too many of these are on display elsewhere except maybe on your dog or cat.

African Trapdoor Spider: This arachnid builds a tunnel with a hinged door made of silk then springs it open to grab its prey.

Gippsland Earthworm: Ever meet a 12' long earthworm? This Australian native is the world's largest.

After going buggy here, step inside the **Mary Ann Lee Butterfly Wing**, a 2,827-foot-long geodesic dome where more than butterflies fly. Look for green and blue dragonflies and black-winged damselflies hovering over the pool. Aquatic bugs like Water Striders inhabit the pool while katydids jump and fly among the shrubs and trees.

The tropical butterfly dome features 300 to 400 flying flowers representing about 33 species of native and exotic butterflies and moths in a lushly landscaped wing, with rock outcroppings and a waterfall. Look for Blue Pansies, Blue Morphos, White Peacocks, Tailed Jays, Great Mormons, Zebra Swallowtails, Great Orange Tips, Malay Lacewings, Dark Blue Tigers and Luna and Cecropia Moths.

Be sure to visit the **Missouri Meadow**, an outdoor garden near the butterfly dome. The peaceful place is a quiet spot to stop and view native nature including butterflies that are attracted to the flowering flora. A wild prairie area overlooking a lake is studded with wildflowers that provide nectar for adult butterflies and food for the caterpillars.

Hours: 9 a.m. to 5 p.m. daily. Closed Christmas Day and New Year's Day.

Admission: (Main Zoo) Free. (Insectarium) $4 per person. Free from 9 a.m. to 10 a.m.

Parking: $7 per car.

Sophia M. Sachs Butterfly House and Education Center

Faust Park

15193 Olive Blvd., Chesterfield, MO 63107
636-530-0076
www.butterflyhouse.org

Walk into a crystal cloud filled with stunning rock formations, lush greenery and flowing flowers and you'll be inside this 8,000-square-foot glass butterfly conservatory. About 1,500 exotic winged beauties from around the world flutter daily through the tropical wonderland. You can see up to 60 different butterfly species like Ruby Lacewings from Malaysia, Goldbanded Foresters from Kenya and electric Blue Morphos from Costa Rica. Discover the Miracle of Metamorphosis display and watch as wet wings emerge from their chrysalides.

An additional 8,000 square feet houses the **Education Center** where arthropod exhibits change every three months or so. Two past buggy exhibits included arachnids from around the world with more than 100 arachnid species and a gut-busting display of entomophagy, or bug eating. Look for the invertebrate exhibit that includes 16 kiosks of creepy crawlies from grasshoppers to ant colonies to a cockroach collection.

Watch a 20-minute butterfly video in the **Emerson Electric Family Theater**, take a buggy class in the **Lopata Learning Lab** or go on a virtual tour of butterflies worldwide on the computers at the **Whitaker Resource Center**.

The backyard of the **Butterfly House** is a **Native Habitat** landscaped with 130 domestic plant species that attract local butterflies. At the front entrance look for the butterfly shaped stones with messages from St. Louis families that pave the way to the house. The **Sculpture Gardens** feature a 30-foot-long Lopatapillar that kids can climb on and a giant Mighty Monarch with faceted eyes. Nectar producing plants surround the sculpture.

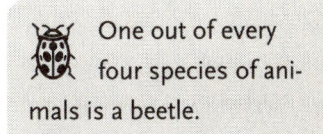

> One out of every four species of animals is a beetle.

Hours: 9 a.m. to 5 p.m. daily, Memorial Day to Labor Day; 9 a.m. to 4 p.m. Tue.–Sun., Labor Day to Memorial Day. Closed Mondays (except national Monday holidays), Thanksgiving, Christmas and New Year's Day.

Admission: $5 adults, $4.50 seniors, $4 ages 4–12, free for ages 3 and under.

The Butterfly Place

2400 State Highway 165, Branson, MO 65616
417-332-2231
www.butterflyplace.com

If you're lucky, an iridescent Blue Morpho will land on your shoulder when you stroll through this 9,000-square-foot butterfly habitat. Owner Bill Hill considers being touched by a butterfly a sign of good luck—a Blue Morpho landed on him more than a decade ago and he started **Butterflies in Flight**, a Naples, FL butterfly farm that supplies many of the flying flowers in this glass-enclosed butterfly habitat.

Up to 2,000 butterflies fly free daily and represent more than 50 species from around the world including Tiger and Rose Swallowtails, Zebra Longwings, Red Lacewings, White Tree Nymphs, Monarchs, Painted Ladies and Queens. When possible native Southern Missouri butterflies are featured such as Cloudless Sulphurs, Great Southern Whites and Spicebush Swallowtails.

The butterflies arrive at the aviary in chrysalides and can been seen hatching in the **Emergence Room**. A great moment happens when a cascade of color is released into this emerald green paradise filled with flowering vines and hanging planters. Wide pathways provide easy access for all visitors, including the disabled. Benches, peaceful music and trickling water offer a relaxing way to commune with nature's winged rainbows.

Lepidoptera exhibits are located adjacent to the emergence room and you can watch an introductory video about the life cycle of a Zebra butterfly before you enter. Visit the companion **Butterfly Place** in Westford, MA.

Hours: 10 a.m. to 5 p.m. late March to late October, 10 a.m. to 4 p.m. November to early December.

Admission: $8.95 adults, $6.95 children, 10 percent off adult price for seniors.

Nebraska

The Butterfly Pavilion

Folsom Children's Zoo and Botanical Gardens

1222 South 27th St., Lincoln, NE 68502

402-475-6741

www.lincolnzoo.org

Flutter by the **Butterfly Pavilion** at this small but delightful children's zoo. Introduced in 2001, the outdoor butterfly aviary features hundreds of native North American butterflies. Packed with plenty of nectar plants, the 20' by 60' screened seasonal sensation includes 28 species such as Red Spotted Purples, American Painted Ladies, Easter Tiger Swallowtails, Tawny Emperors, Long Tailed Skippers, Ruddy Daggerwings, Buckeyes, Queens, Julias, Viceroys and Monarchs. Be sure to pick up the colorful brochure that helps you easily identify the bevy of butterflies. Look for the emergence chamber where you can watch the delicate wings unfurl from their chrysalides.

Afterwards, visit the **Animal Kingdom** Building where you'll find the **Bumble Bee Discovery Kiosk**. Use a touch screen at this interactive display to learn more about these pollinators from the importance of bees to Nebraska agriculture to identifying bees. A **University of Nebraska** entomologist who runs a student science group called the **Bumble Boosters** (http://bumbleboosters.unl.edu/) set up the kiosk. The building also houses exhibits of Madagascar Hissing Cockroaches, Vietnamese Walking Sticks and Brown and Chilean Rose Tarantulas on low kid-friendly counters. Look for plenty of other wildlife from Meerkats and Squirrel Monkeys to Bearded Dragons and Fruit Bats.

Just before the butterflies are gone, the zoo puts on a big **Bug Bash!** in September, featuring activities such as roach races, bee wrangling and Monarch tagging (see Bugged Out chapter).

Hours: (Zoo) 10 a.m. to 5 p.m. (8 p.m. on selected Wednesdays) April 15 to October 15 only. (Butterfly Pavilion) Early June to late September.

Admission: $5 adults, $4 seniors, $3 ages 2–11, free for ages under 2.

Discovery Room

Henry Doorly Zoo
3701 S. 10th St., Omaha, NE 68107
402-733-8401
www.omahazoo.com

This kid-friendly space at the zoo's **Wild Kingdom Pavilion** is jumping with a variety of invertebrates and insect interactives. More than a dozen 10-gallon aquariums house walking sticks, Dung Beetles, Praying Mantids, scorpions, millipedes, centipedes, Cave Crickets, grasshoppers, Assassin Bugs, Cactus Beetles, tarantulas, Orb Spiders and Black Widows. A highlight is the Leaf Cutter Ant colony, the fungal farmers of the insect world. A 55-gallon aquarium holds aquatic arthropods like Diving Beetles, Water Scorpions and Giant Water Bugs. During the spring, there's usually a display of caterpillars such as Tobacco Horn Worms, moth larvae or silk worms.

Use the Optech camera to get a close up view of Madagascar Hissing Cockroaches or play interactive games and learn about insects on two computers. Microscopes and sliding magnifiers let you view insect and other animal specimens. Be sure to look at the graphics on the walls that feature the Olympian feats of insects. Bring your camera and take a picture of the kids sticking their heads in a photo board where they become a flower or a bee. In the foyer, look for artist Patrick Bremer's wooden sculptures of a Praying Mantis, wasp and dragonfly.

The pavilion is also home to other animals such as reptiles, amphibians, birds, bats, and small primates from Gila Monsters to hedgehogs to Tree Kangaroos.

An indoor butterfly exhibit is planned for 2004.

Hours: 9:30 a.m. to 5 p.m. daily. Closed Thanksgiving, Christmas, and New Year's Day. Visitors may stay on zoo grounds two hours after closing or until dusk.

Admission: $8.50 adults, $7 seniors (62+), $4.75 ages 5–11.

Nevada

Las Vegas Natural History Museum

900 Las Vegas Blvd. North, Las Vegas, NV 89101
702-384-DINO
www.lvnhm.org

Even in the grownup desert play land of Las Vegas, there's a kid-friendly place to see live bugs. This 30,000-square-foot multi-sensory museum houses a host of live animals from pythons to geckos to hedgehogs. Most of the critters in the **Wild Nevada Room** can all be found within 100 miles of Las Vegas, including some of the invertebrates: Emperor, Desert Hairy and Flat Rock Scorpions, Rose Hair and Cinnamon Brown Tarantulas, a Black Widow and Funnel Web Spider, a couple of Giant Madagascar Millipedes and a colony of Cuban Cockroaches. The museum also holds a variety of hands-on weekend events, including some on bugs and butterflies.

Be sure to visit the museum's cool blue and green **Marine Life Room** with small live sharks in a 3,000-gallon tank, the **Young Scientist Center** with interactive displays and a **Learning Center** where kids can dissect owl pellets. Look for the big robotic dinosaurs and the collections of fossilized and mounted animals.

Hours: 9 a.m. to 4 p.m. daily. Closed Thanksgiving and Christmas.

Admission: $5.50 adults, $4.50 seniors, $3 ages 4 to 12.

New Hampshire

Squam Lakes Natural Science Center

Route 113, Holderness, NH 03245
603-968-7194
www.nhnature.org

Buzz with bees, boogey with butterflies and get antsy with ants at this 200-acre science center with its unique combination of outdoor and hands-on exhibits. It's probably the only place in the state to see live bugs.

A walking trail connects outdoor exhibits with exhibit buildings. About three-quarters of the way along the main exhibit loop you'll find the Honey Bees humming in an observation hive in the bear exhibit. The Carpenter Ants are working on their farm in the soil exhibit. The ants might be off exhibit when you visit due to renovation of the display.

The live shimmering stars of the science center are the native butterflies that flock to the one-acre **Kirkwood Gardens**, named after local landscaper Sunny Kirkwood. Formerly an old gravel parking lot, dozens of volunteers, the New Hampshire Landscape Association and local nurseries pitched in to produce a garden attractive to butterflies and bees. Look for Monarchs, Tiger Swallowtails, Painted Ladies, Sulphurs, Fritillaries, Cabbage Whites, Common Blues, and Hummingbird Moths zipping through perennials such as Yarrow, Honeysuckle, Cosmos, Daisies and Asters. Plant sales are held to keep the garden growing.

Free and open to the public at all times, the gardens can be accessed off Route 3 or through the center's exhibit trail.

Hours: 9:30 a.m. to 4:30 a.m. daily May 1 to November 1.
Admission: $9 adults, $6 ages 5–15, free for ages 4 and under, July

and August. $7 adults, $4 ages 5–15, free for ages 4 and under, May, June, September and October.

New Jersey

Philadelphia Eagles Four Seasons Butterfly House and Education Centre

Camden Children's Garden

3 Riverside Dr., Camden, NJ 08103
856-365-8733
www.camdenchildrensgarden.org

Visiting the only children's garden in the Garden State is like stepping onto the pages of *The Secret Garden* or *Alice in Wonderland*, both of which have themed gardens named after them. The 4.5-acre interactive garden is part of the non-profit **Camden City Garden Club** and its newly opened year-round **Butterfly House and Education Centre** treats everyone to fluttering bouquets of butterflies.

About two hundred butterflies bask in the 1,200-square-foot glass greenhouse, featuring a peaceful pond, bubbling fountains and soft music. Located next to the **Potting Shed** you can't miss its gigantic Noah's Arc façade. More than seven species are raised in a butterfly nursery and released inside to waft among the Lantana, Pentas, Marigolds, Salvia, Mints, Verbena and Heliotrope. Look for the plants in pots and hanging baskets hung at every level in the 15-foot-high house so that butterflies that fly at different heights can nourish themselves on the nectar-bearing flora. Look for Monarchs and Painted Ladies at a variety of heights, Pipevine and Black Swallowtails at mid-range height and Zebra Longwings and Julias at the higher elevations. Red Admirals are harder to spot but look for them laying eggs on Stinging Nettle, their favorite host plant.

Host plants are plentiful in the house but are always having to be replaced because the caterpillars begin to eat them of course! Look for the eggs and caterpillars of Monarchs on the Tropical Milkweed, Painted Ladies on Mallow or Hibiscus, Pipevine Swallowtails on Tropical Pipevine while their cousins, Black Swallowtails, prefer herbs such as Parsley, Dill and Fennel. You'll see Julias and Zebra Longwings depositing their eggs while flying to the tips of Passion Flower Vines. Newly hatched caterpillars don't stay in the house for long; they are gently removed and brought to a nursery where they can contently munch.

Be sure to visit the 1/2-acre **Butterfly Garden** and **Butterfly Stop** outside the enclosed butterfly house. You know you're there when you see the big statue of writer and poet Walt Whitman holding a butterfly on his finger. The carousel is here too. Butterfly-friendly plants are in abundance: Russian Sage, Milkweed, Butterfly Bush, Rue, Yellow-Twigged Dogwood, Wild Rose and Tansy.

Stop at the **Picnic Garden** near the butterfly areas where kids can experience what it's like to be ants marching across a picnic blanket complete with cups, bowls and a picnic basket. Everything is larger than life from the oversized red and white painted picnic blanket to the giant picnic basket that's actually a grape harbor. The ants are made of car parts and are people size. Strawberry gardens and blueberry bushes fill the teacups that also serve as great hiding places for kids. The large plates are planted with vegetable gardens, varieties of tomato and herbs.

Hours: (Garden) 9:30 a.m. to 5:30 p.m. April 16 to September 15; 9:30 a.m. to 4:30 p.m. September 16 until April 15; 10 a.m. to 5 p.m. weekends. (Butterfly House) 10 a.m. to 1 p.m., 2 p.m. to 4 p.m. weekdays, 10 a.m. to 4 p.m. weekends.

Admission: $5 adults, $3 children.

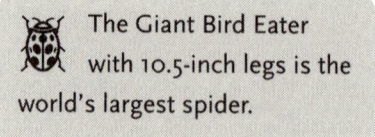

> The Giant Bird Eater with 10.5-inch legs is the world's largest spider.

Kate Gorrie Butterfly House

Buttinger Nature Center

Stony Brook-Millstone Watershed Association
31 Titus Mill Rd., Pennington, NJ 08534
609-737-7592
www.thewatershed.org

The largest private space in central New Jersey, this 785-acre reserve and nature center is home to a natural butterfly house. Open to the elements, the 1,500-square-foot mesh structure's butterfly garden beckons a variety of local species to flutter inside. From spring until fall the house provides food and shelter for butterflies and their larva.

Look for Eastern Tailed Blues, Summer Azures, Gray Hairstreaks, American Coppers, Clouded Sulphurs, Black Swallowtails, Great Spangled Fritillaries, Red Admirals, Painted Ladies, Monarchs, Little Wood Satyrs, Juvenal's Duskwings, Silver Spotted Skippers, Pearl Crescents, Eastern Commas and Question Marks. Learn how to identify some of these natives then go on a butterfly scavenger hunt in the house for eggs, caterpillars and chrysalides.

The reserve also offers more than eight miles of trails through fields, forests, streams and wetlands. Look for more butterflies here, as well as other wildlife. There are wonderful wildflower trails, a research pond and some of the oldest trees in New Jersey. Be sure to visit the **Discovery Room** inside the **Buttinger Environmental Education Center** as well as the **Watershed's** organic farm.

Hours: (Butterfly House) Daylight hours May to October.
Admission: Donation

Bug Zoo

Liberty Science Center

Liberty State Park, 251 Phillip St., Jersey City, NJ 07305
201-200-1000
www.lsc.org

Most science centers, unlike nature centers, don't have many live bugs let alone an opportunity to pet one. **Liberty Science Center** does. Head for the center's **Environmental Floor** where you get sweeping views of the Hudson River, New York City and the Statue of Liberty. Here in **Micro Viewpoints**, use microscopes to magnify tiny creatures called microbes (mini bugs) and without scientific equipment, get up close to dozens of local and exotic invertebrates in the **Bug Zoo**.

Housed in terrariums with a 360-degree view, are Central American Giant Cave Cockroaches, African Millipedes from Kenya, tarantulas and Giant Prickly Insects. Since the Honey Bee is New Jersey's state insect, there's also a 10-foot-wide full working beehive that you can observe from both sides. Look for the portals that connect the hive to the outside world.

Want to pet a Madagascar Hissing Cockroach? Then enter the **Greenhouse Discover Room** on the same floor where bug wranglers let you touch the roach's shiny exoskeleton. Joining the roaches are a Flat Rock Scorpion, Pink Toe Tarantula, Arizona Tarantula and a colony of American Cockroaches. The room also features several snakes and reptiles, some of which are brought out for show and tell.

You'll find the cousins of the land invertebrates in **The Estuary** exhibit on the same floor. Explore the different types of aquatic life and get a closer look at water invertebrates in the **Touch Tank**: Horseshoe Crab, Sea Urchin and a Nine-Spined Spider Crab.

Hours: 9:30 a.m. to 5:30 p.m. daily. Closed Thanksgiving Day and Christmas Day.

Admission: $10 adults, $8 seniors (62+) and ages 2–18, $2 teachers with I.D., free for ages under 2.

New Mexico

PNM Butterfly House

Rio Grande Botanic Garden

Albuquerque Biological Park
2601 Central Ave. NW, Albuquerque, NM 87104
505-764-6200
www.cabq.gov/biopark

More than 500 North American flying flowers flutter through this wood and mesh butterfly house that sports a design that is almost as colorful as the butterflies inside. Linemen from Power New Mexico erected the 40 wooden poles that serve as the basic structure for the 3,392-square-foot facility that's topped with vinyl mesh covered panels of red, purple, yellow and blue that tilt and provide slight shade while letting in the sun's rays.

Look for Zebra Longwings, Anise Swallowtails, California Sisters, Malachites, Question Marks, Sara Orangetips, Anemone Fritillaries and Crimson Patches weaving through the 3,000 blooming plants. About 200 species of flora are used to create the lush landscape necessary for more than 30 species of butterflies to survive. Sleepy Oranges, White Peacocks, Red Admirals, Queens and Giant Swallowtails also dance among the Hollyhocks, Butterfly Bush, Black-Eyed Susans and Purple Coneflowers planted inside.

Outside the pavilion a nectar garden filled with Shasta Daisies, English Lavender, Blue Catmint and dozens of other plants attracts local

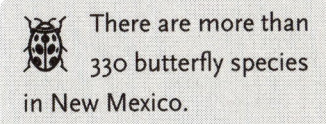

> There are more than 330 butterfly species in New Mexico.

butterflies and hummingbirds. Be sure to look for the "brooder" room inside the pavilion, where butterflies emerge from their chrysalides and unfurl their wings for the first time.

Before you leave, make a special visit to the **Children's Fantasy Garden** where kids are treated to a larger than life view of gardening horticulture. Look for the huge rabbit hole where six-foot earthworms burrow through the walls, an 11-foot-tall watering can, a two-story walk-through pumpkin that's a whopping 42 feet in diameter plus a garden full of gargantuan veggies.

Part of a larger **BioPark**, admission to the botanic garden includes the aquarium.

Hours: (Garden) 9 a.m. to 5 p.m. Tue.–Sun.; 9am–6pm Sat.–Sun., June through August. (Butterfly House) May to September. Closed Mondays (including Monday holidays), Thanksgiving, Christmas and New Year's Days.

Admission: $5 adults, $3 seniors (65+) and ages 3–12.

Frances V.R. Seebe Tropical America

Rio Grande Zoo

**Albuquerque Biological Park,
903 Tenth St. SW, Albuquerque, NM 87102
505-764-6200
www.cabq.gov/biopark**

Tarantulas are trolling the grounds of the zoo's new tropical building. Actually, the several species of spiders live in terrariums embedded in an artificial rock wall that's part of this immersion exhibit. Walk through the jungle mist to find Pink Toes from South America, Costa Rican Zebra Tarantulas, Mexican Redrumps and Curly Hair Tarantulas. Joining the Big Ts are Spider Monkeys, Cotton Top Tamarins, toucans, Prehensile-Tailed Porcupines and boas.

During the summer the zoo opens up its **Kaleidoscope Discovery**

Room where invertebrates are the main attraction. Meet Vinegaroons, millipedes, tarantulas and if you dare, pet a Madagascar Hissing Cockroach. Hermit Crabs and snails are sometimes on display and teen volunteers often bring in local bugs for a short-term display. Use the Bioscanner to get close up looks at the aquatic bugs on display and look through microscopes to magnify insect specimens. Kids can put on impromptu puppet shows with a wide variety of animals including bugs or make crafts at the activity tables.

Hours: 9 a.m. to 5 p.m. Tue.–Sun., 9 a.m. to 6 p.m. Sat.–Sun., June through August. Closed Mondays, Thanksgiving, Christmas and New Year's Day.

Admission: $5 adults, $3 seniors (65+), and ages 3–12.

New York

The Butterfly Conservatory: Tropical Butterflies Alive in Winter

American Museum of Natural History

Central Park West at 79th St., New York, NY 10024
212-769-5100 (info) 212-769-5200 (tickets)
www.amnh.org

Step in from the cold sidewalks of New York and into a warm tropical environment surrounded by lush foliage and shimmering rainbows of live brilliant butterflies. It can happen and does during the winter months at this prestigious museum.

A 1,300-square-foot translucent vivarium is resurrected every October and stays open until late May. Stroll through a jungle of plants and trees in this freestanding structure and watch as 500 tropical butterflies dance around the vibrant blossoms. Look for electric Blue Morphos, Giant Swallowtails, Isabella Tigers, Zebra Longwings, Malachites and Paper Kites. Powerful Halide lamps simulate streams of sunlight encouraging Scarlet Swallowtails, Tailed Jays, Banded Oranges, Queens and Clippers to fly freely. Visitors standing outside can peer inside and watch as Julias, Owls, Buckeyes and Question Marks flutter among the strolling visitors.

More than 50 species of butterflies call this home, with weekly shipments of 500 chrysalides from butterfly farms in Florida, Texas, Costa Rica and Africa keeping the population stable. Some of the chrysalides hang in a case inside the conservatory so you can see the adults emerge. A brochure with color photos of each butterfly helps you identify these winged beauties.

Learn about butterfly biology through the educational displays outside the conservatory. Discover how designs on a butterfly's wings are formed, how some butterflies protect themselves from predators and how scientists use wild butterflies to measure the health of an ecosystem. On video screens you can watch a half-hour BBC documentary, *The Butterfly: Beauty or the Beast?*

Back home you can revisit the winter conservatory via the Internet. A digital camera captures the Lacewings, Painted Ladies, White Peacocks and Eastern Tiger Swallowtails and transmits the images to the museum's Web site.

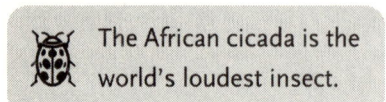

> The African cicada is the world's loudest insect.

A warning: this exhibit is so popular that timed tickets are given out. You can reserve them in advance, however. Since the "weather" is much warmer inside the vivarium than outside the museum, you may want to deposit coats and hats at the Coat Checks.

Hours: (Museum) 10 a.m. to 5:45 p.m. daily. Closed Thanksgiving and Christmas. (Butterfly Conservatory) October to May. Last timed entry is 5 p.m.

Admission: $15 adults, $11 students and seniors, $9 ages 12 and under, $7.50 members, includes museum admission.

Butterfly Zone

Bronx Zoo

Fordham Rd. and the Bronx River Parkway, Bronx, NY 10460
718-367-1010
www.wcs.org

Ever walk into the mouth of a giant caterpillar? You will, when you enter one of the largest butterfly enclosures ever built. This astounding temporary exhibit uses mesh and steel hoops to create the 170-foot-long shape of a butterfly larva.

Located on the **Astor Court Lawn**, the 42-foot-wide award-winning butterfly house features more than 35 species of native butterflies such as Great Purple Hairstreaks, Red Spotted Purples, Silver Spotted Skippers, Hackberries, Tawny Emperors, White Admirals, Variegated Fritillaries, Zebra Longwings, Ruddy Daggerwings, Clouded and Cloudless Sulphurs. Eight species of the large and lovely Swallowtail shimmer their wings here: Pipevines, Polydamas, Zebras, Giants, Eastern Tigers, Palamedes, Eastern Blacks and Spicebushes. There are also several species of moths including Lunas, Ios, Hummingbirds and Carolina Sphinxes.

More than 3,000 plants of 75 different species provide nourishing nectar for the adult butterflies and plenty of food for the caterpillars. Look for 22-foot-tall Golden Rain Trees, Sunflowers, Black-Eyed Susans, six Milkweed species, Spicebush, Pipevine and hundreds of flowering flora. The **Forest Edge** area of the exhibit re-creates a butterfly garden that you'd find in the Deep South using trees and shrubs such as Pigeon Berry and Lantana to attract Giant Swallowtails, Julias, Question Marks and Commas. The **Meadow** area of the exhibit is planted to simulate a local wildflower meadow with Daises, Butterfly Bush and Butterfly Weed to attract Monarchs, Painted Ladies and Tiger Swallowtails.

Outside the exhibit, another flowering garden surrounds the giant caterpillar. The plants are neatly identified so visitors can learn what butterfly-attracting and caterpillar-friendly plants they can put in their own backyard.

The exhibit includes two more fun aspects: photo opportunities that let you pose as a big butterfly hovering over a field of smiling daisies or as a cute caterpillar lounging on a leaf. At the end of the enclosure you'll find the 900-square-foot gift shop bulging with a bounty of butterfly items.

Hours: (Butterfly Zone) Memorial Weekend until October 1. Call for summer hours.

Admission: (Butterfly Zone) $2 per person. Call for zoo rates.

Central Park Zoo

830 Fifth Ave., New York, NY 10021
212-861-6030
www.wcs.org

Step off Fifth Avenue into **Central Park** and you'll discover the Manhattan jewel of the **Wildlife Conservation Society**. Transformed by the society in the 1980s, the zoo has become an urban paradise with a tropical oasis tucked in the corner. The steamy **Rain Forest** exhibit is home to tropical birds, rare monkeys, roaming reptiles and amphibians. A highlight of the forest is the Leaf Cutter Ant Colony.

Remember the ants in Disney's *A Bug's Life* film? These are the same critters, except they're real ones, not animated, and they're very busy farming their fungal garden not fighting giant crickets. The colony of Atta cephalotes is just past the **Mouse Deer/Riverbank** exhibit and the bat cave. Revamped in 2001, the Leaf Cutter Ant display is more accessible than before.

See the ants gather leaf cuttings in a 12' square forage area, then follow them through a mound with an entrance hole and a cut away view of 15 chambers and interconnected tunnels. Macro digital cameras installed in the fungus farm let you view the Leaf Cutters on flat panel monitors overhead. Watch as workers tend the garden, carry leaves and clean house. There's also a time-lapse short film of the fungal garden's growth over a two-week time span shown on two separate monitors. A large backlit graphic gives you the background on Leaf Cutter life.

Look for Madagascar Hissing Cockroaches and tarantulas in the **Rain Forest**, too. For the ocean cousins of these land lubbers, look in the large cold water reef tank in the **Penguin Building** where sea anemones and limpets live. The zoo has plans to expand its invertebrate collection in the near future.

If you visit in the spring, look for the "butterfly garden" on the west hill near the **River Otter** exhibit. The horticulture department puts extra butterfly attracting plants in place so local species of flying flowers and migrant birds will pay a particular visit to the zoo and its 1,400 animals and new **Tisch Children's Zoo.**

Hours: 10 a.m. to 5 p.m. Mon.–Fri., 10 a.m. to 5 p.m. Sat.–Sun. & holidays, year-round.

Admission: $3.50 adults, $1.25 seniors (65+), 50 cents ages 3–12.

Tropical Forest

Staten Island Zoo
614 Broadway, Staten Island, NY 10310
718-442-3100
www.statenislandzoo.org

The invertebrates aren't numerous here and are a little hidden away, but where they are is an incredible place to visit. The **Ralph J. Lamberti Tropical Forest** is a multi-million dollar re-creation of an endangered South American tropical forest. Inside this natural habitat filled with flora and fauna are Conyers, Tamarins, Iguanas, Grey Winged Trumpeters and Sloths. The animal enclosures are woven into the fabric of the forest and among them you'll find a collection of about 20 aquariums sunk into two walls and inhabited by terrestrial invertebrates. Look for Emperor Scorpions, Giant Centipedes, Assassin Bugs, Giant African Millipedes, Vietnamese Walking Sticks, Giant Bird Eaters, Pink Toe Tarantulas and Black Widows. There's also a collection of live marine invertebrates such as corals and Sea Anemones. The zoo also features an aquarium, so if you're in the mood to see more spineless wonders, it's adjacent to the

Tropical Forest building.

Hours: 10 a.m. to 4:45 p.m. daily. Closed Thanksgiving, Christmas and New Year's Day.

Admission: $3 adults, $2 ages 3–12, free for ages under 3.

Bugs and Other Insects

Staten Island Children's Museum

Snug Harbor Cultural Center
1000 Richmond Terrace, Staten Island, NY 10301
718-273-2060

"Don't squish 'em...study 'em" is the motto at this popular museum where the **Bugs and Other Insects** exhibit let's kids creep and crawl in their exploration of these creatures. Kids get a bug's eye view of life as an insect when they crawl through a larger-than-life model ant colony with grass topped tunnels filled with chicken-egg-sized ant eggs. In **Take a Closer Look** kids can turn several magnifying cylinders containing insects while in **Create A Critter** kids can take the three main body parts of different insects and mix and match them to form new creatures. In **Whose Baby Am I?** kids use clues to properly pair the description of an adult insect with the description of the baby bug. **Where are My Ears?** uses lift-the-flaps to show how a Southwestern Grasshopper hears and eats. At four rubbing stations kids can use crayons and paper to bring out images of a dragonfly, beetle, house fly or butterfly.

Art plays a special bug role here, with two large murals by Julia Healy depicting common indoor and outdoor scenes. Kids can lift flaps to reveal live insects that would ordinarily be found in a pond or under the kitchen sink. There are also several display cases of more than 200 bug specimens for kids to examine.

And what about live bugs? The museum has them too. There are Emperor Scorpions, Giant African Millipedes, Hermit Crabs, a Chilean Rose Hair Tarantula, crickets, Madagascar Hissing Cockroaches and during the warmer months, Chinese Praying Mantids. It's also during the spring and summer that the museum builds a small butterfly cage so kids can see Painted Lady Butterflies go through their metamorphosis. The butterfly's Milkweed-eating caterpillars are here too along with black and red Milkweed Bugs, one of the species of invertebrates known as "true bugs." You can see the chrysalides on display too, and on a good day, witness an emerging butterfly.

Hours: Noon to 5 p.m. during school year; 11 a.m. to 5 p.m. during summer. Closed Mondays, open most school holidays.

Admission: $4 ages 2 and older, free for members and teachers.

Honey Bee Hive and Ant Colony

Sciencenter

601 First St., Ithaca, NY 14850
607-272-0600
www.sciencenter.org

Follow the marching Leaf Cutter Ants as they haul their parasols to their nest on the first floor near the front entrance. The parasols are actually leaves that are more than twice their size that the ants chew up and use to grow fungus. The ants in this mound hail from the island country of Trinidad and you can get a closer look at them with the science center's magnifying glasses. Be sure to look for more information on these active ants in the display's accompanying bug book.

Another social bug colony, a Honey Bee hive, is located on the second floor. The 2,000 domestic bees provided by the Finger Lakes Beekeepers Association, live in an observation house on the museum's second floor. Watch them buzz in and out of their Plexiglas home through a tube that leads to the outside. Learn how to do the Bee Dance when center volunteers and staff members demonstrate how bees communicate with each other. Be sure to read more about these

> The world's heaviest insect is the African Goliath Beetle at 1/4 lb.

honey makers in the book attached to the display.

Hours: 10 a.m. to 5 p.m. Tue.–Sat., Noon to 5 p.m. Sun.

Admission: $4.50 adults, $4 seniors, $3.50 ages 3–12, free for ages under 3.

Native Species Butterfly House

Visitor Interpretive Center

Adirondack Park, Route 30, Paul Smiths, NY 12970
518-327-3000
www.northnet.org/adirondackvic/

Only a half hour from Lake Placid is a six-million-acre park with 2,000 miles of hiking trails and the highest peaks in New York State. The park has two terrific wildlife interpretive centers and at this one you'll find the butterfly house that set the standard for other native species houses.

The 30' by 50' seasonal structure flutters with butterflies and moths that have been collected from the park such as Bronze and Bog Coppers, Aphrodites, Meadow Fritillaries, Silvery Checkerspots, Green Commas, and Ringlets. The intimate setting of the small enclosure offers visitors a chance to view these beauties up close. As you walk along the U-shaped path lined with hundreds of nectar and host plants you may see Tawny Crescents, Northern Pearly Eyes, Columbine Dusky Wings, Long Dashes and Pepper and Salt Skippers.

View butterflies emerging from a chrysalis and look for eggs and caterpillars. Can't decide whether you're looking at a Black Swallowtail or a Spicebush Swallowtail? Then ask the **Adirondack Park Institute** Butterfly Naturalist or trained volunteers who are here to help. Look for the prepared checklist of the more than 75 species of butterflies you may encounter at the park.

Snow can fall early and late in these mountains so the house is only open for a limited time. While the butterfly house is free, it relies on member fees and donations to supply the plants needed to keep the butterflies happy. Be sure to ask a naturalist how to find the bug-eating plants in the bog on the **Forest Ecology Trail.**

Hours: (Park) 9 a.m. to 5 p.m. daily. Closed Thanksgiving and Christmas. (Butterfly House) 10 a.m. to 4 p.m. daily mid-June until Labor Day.

Admission: Free.

North Carolina

Magic Wings Butterfly House & Aventis CropScience Insectarium

Museum of Life and Science

433 Murray Ave., Durham, NC 27704

919-220-5429

www.ncmls.org

Life sciences really come to life at this 70-acre indoor/outdoor museum, especially when you're talking about butterflies and bugs. Part of a larger **BioQuest** project, the year-round **Magic Wings Butterfly House** hums with the sound of 1,000 exotic butterflies fluttering. The 5,000-square-foot conservatory is the largest museum butterfly house in the Southeast and features Blue Morphos from Ecuador, Scarlet Mormons from the Philippines, Mother of Pearls from Kenya, Owls from Peru, Crimson-Patch Longwings from Costa Rica and Paper Kites from Malaysia.

> Only adult insects have wings.

About 50 species may be flying through the three-story glass conservatory at any one time. The chrysalides are imported from their countries of origin and released daily as they emerge. Watch the magic of metamorphosis in the **Mary Martha Uzzle Emerging Wonders Room** where you can see hundreds of adult butterflies first unfurl their damp wings and learn about the butterfly life cycle by observing real butterfly eggs and larva.

As you walk though the tropical conservatory in the 80-degree/80 percent humidity environment, you can't help but notice its exotic botanicals. There are more than 250 species of rare and tropical plants. Look for Cacao Trees (where chocolate comes from), Firebush, Vanilla Orchids, Coffee Trees, Cinnamon, Papaya Trees, Frangipani, Ginger, carnivorous Tropical Pitcher plants and the Red Orchid tree, of which only 50 exist in the world.

The exotics aren't the only rainbows flying here, at least during the warmer months of the year. The 18,000-square-foot **Magic Wings** building features the seasonal **Carolina Pavilion** with up to 25 different North American species dancing through the landscape. From Memorial Day to early October watch as these butterfly families entertain you: Zebra Swallowtails, Monarchs, Gulf Fritillaries, Zebra Longwings and several moths. Hundreds of butterflies dazzle you with their bright colors under the semi-shade cloth. Be sure to wear yellow, their favorite color, and they may land on your shoulder.

Hear the crunch and munch of a caterpillar eating at the 1,700-square-foot insectarium where the buzzing of bugs is more than audible. Use specialized video and audio equipment to see and hear more than 25 exotic and native species of insects do their thing. Get close to live bugs such as Giant Walking Sticks; Death's Head Cockroaches with heads resembling skulls; Goliath Bird Eaters, the largest spiders on Earth; Dead Leaf Mantids which resemble their name and Giant Orb Weaver Spiders which roam freely but safely on the insectarium's ceiling corners. There's also a giant farm of Harvester Ants, native insects that harvest seeds.

Hours: 10 a.m. to 5 p.m. Mon.–Sat., Noon to 5 p.m. Sun., daily; until 6 p.m. Fri.–Sun. Memorial Day through Labor Day. Closed Thanksgiving, Christmas and New Year's Day.

Admission: $8.50 adults, $7.50 seniors (65+), $6 ages 3–12, free for ages 2 and under.

Butterfly Pavilion

Charlotte Nature Museum

1658 Sterling Rd., Charlotte, NC 28202

704-372-6261

www.discoveryplace.org

Hundreds of North American butterflies wave their wings in this 1,000-square-foot glass atrium overlooking the museum's nature trail. Monarchs, Queens, Zebra Longwings, Julias, several Swallowtail species and Crimson Patch Butterflies loop through the flowerbeds. Learn about the butterfly life cycle and watch these beauties emerge from their chrysalides in the **Butterfly Foyer**. This area introduces you to the different butterflies you'll see in addition to some live invertebrates and mounted specimens.

Dedicated to wildlife of urban North Carolina, the museum is associated with a larger science center, **Discovery Place**, also in Charlotte. The museum's quarter-mile-long **Paw Paw Nature Trail** below the butterfly pavilion features a wildflower garden and a large stand of native Paw-Paw trees. Cross **Little Sugar Creek** on the suspension footbridge and you enter **Freedom Park** and lake. Look for geese, ducks, herons, frogs and turtles.

Hours: 9 a.m. to 5 p.m. Mon.- Fri., 10 a.m. to 5 p.m. Sat., 1 p.m. to 5 p.m. Sun. Closed Thanksgiving, Christmas Eve, Christmas Day, New Year's Day and Easter Sunday.

Admission: $4 per person ages 3 and up.

Knight Rain Forest

Discovery Place

301 North Tryon St., Charlotte, NC 28202

704-373-6261

www.discoveryplace.org

Crawling around this three-story moist tropical habitat are Madagascar Hissing Cockroaches, hairy tarantulas from South America and giant millipedes. Well, not literally. These invertebrates are brought out for you to see and sometimes touch during daily hands-on programs in the **Rain Forest Theatre**.

Step on a balcony and discover the other animals living among the Banana Trees such as the Blackneck Aracari, a member of the Toucan family; Burmese Pythons, Green Iguanas and Red-Footed Tortoises. For more creature close encounters, be sure to visit the **Discovery Place's Charlotte Nature Museums** and its **Butterfly Pavilion**.

Hours: 9 a.m. to 6 p.m. Mon.- Thu., 9 a.m. to 6 p.m. Fri.–Sat., 1 p.m. to 6 p.m. Sun., June to Labor Day. 9 a.m. to 5 p.m. Mon.–Fri., 9 a.m. to 6 p.m. Sat., 1 p.m. to 6 p.m. Sun., September to May.

Admission: $7.50 adults, $6.50 seniors (60+), $6 ages 6–12, $5 ages 3–5, free for ages under 3.

Living Conservatory & Arthropod Zoo

The North Carolina Museum of Natural Sciences

11 West Jones St., Raleigh, NC 27601
919-733-7450
www.naturalsciences.org

The Southeast's largest natural history museum hosts an unusual habitat house on its fourth floor. A re-created *dry* tropical forest, a habitat more endangered than the rain forest, you'll know you've reached the **Living Conservatory** from the aroma of live Vanilla plants. Living color in the form of shimmering Blue Morpho Butterflies and Ruby-Throated Hummingbirds zip through the live Vanilla Orchids, Wild Yams and dazzling Red Hot Lips plants.

A real 30-pound Two-Toed Sloth lounges on a fabricated Strangler Fig tree in the 1,300-square-foot habitat. Ornate forest turtles burrow in the soil around the tropical plants and Stewart's Milk Snakes slither around the rock outcrop and pond. Weaving among these live creatures are 10 to 12 different butterfly species including Orange Longwings, Giant Swallowtails, Monarchs, Julias, Owls, Zebra Longwings and Postman Butterflies.

The **Living Conservatory** is a guided walk-through experience so visitors are admitted in groups of 15 to 20 people. Due to staffing fluctuations the times the conservatory is open may vary, so call ahead.

Bugs and other jointed-legged creatures with exoskeletons can be found in the **Arthropod Zoo**, adjacent to the conservatory. You can't miss it. An eight-foot tall model of a Praying Mantis and a dragonfly model 8,000 times larger than life size serve as sentries outside the entrance to the 2,860-square-foot exhibit. The changing cast of characters includes a Giant Bird Eater Tarantula, Madagascar Hissing Cockroaches, Hercules Beetles, a Giant African Millipede, Giant Walking Sticks, a Giant Centipede, Black Widow, barnacles, lobsters, crayfish and crabs including a scary looking Horseshoe Crab. Trained handlers bring out the safe bugs for visitors to touch daily.

There are plenty of interactives and oversized models to keep your attention. See the world through the compound eyes of a Honey Bee, push buttons to hear the sounds that a Conehead Katydid or a Dogbane Tiger Moth makes. Buzz through bug locomotion when you watch a high-energy video and see *What's for Dinner?* with bugs displayed on the plates of people from other cultures.

The museum also offers summer camps with bug and butterfly themes as well as buggy field trips for adults. Be sure to check out **BuGFest!** held every July. The museum wriggles with activity with Roach Races, beehive demonstrations, butterfly gardening tips, bug crafts and the Café Insecta serves up free samples of bug cuisine (see Bugged Out chapter).

Hours: (Living Conservatory) Noon to 4 p.m. Tue.–Sat., 1 p.m. to 4 p.m. Sun. (Museum) 9 a.m. to 5 p.m. Mon.–Sat., Noon to 5 p.m. Sun. First Fri. of every month (except January) the museum stays open until 9 p.m. Closed Thanksgiving Day, Christmas Eve and Christmas Day. Living Conservatory, Discovery Room, and Naturalist Center closed Mondays.

Admission: Free. Some charges for special exhibits. Children 13 and under must be accompanied by an adult.

Naturalist Center

Catawba Science Center

243 Third Ave. NE, Hickory, NC 28603
828-322-8169
www.catawbascience.org

The imposing brick building in the **Arts & Science Center** of **The Catawba Valley** may not seem a likely locale for live creepy crawlies. But step inside and you'll change your mind. The center features a bevy of bugs including Rose Hair Tarantulas, Georgia Millipedes, Giant Walking Sticks from Papua New Guinea, Madagascar Hissing Cockroaches and Emperor Scorpions. Home to live reptiles and amphibians, the center even has a small South American Alligator that was dumped in the North Carolina mountains. At the **Backyard Beasts** exhibit, take a close look at a variety of native critters by operating a video camera with a joystick. The center also has dozens of preserved animals and plenty of microscopes, magnifiers and nature identification field books.

Hours: 10 a.m. to 5 p.m. Tue.–Fri., 10 a.m. to 4 p.m. Sat., 1 p.m. to 4 p.m. Sun. Closed Mondays and major holidays.

Admission: $4 adults, $2 seniors (62+) and ages 3–16, free for ages under 3.

North Dakota

Discovery Center

Dakota Zoo

Sertoma Park, Riverside Park Rd., Bismark, ND 58502
701-223-7543
www.ndtourism.com/regions/west/WestZoo.html

What may be the only display of Thatching Ants in the U.S. is raising the roof at North Dakota's largest zoo. The colony of 10,000 has created a thatch mound about three feet high inside the 6,300-foot **Discovery Center** that's designed to look like a wood barn. Watch the ants in their two-foot-square exhibit as they tear up thatching material, make piles and crawl through tunnels.

Get up close to a bounty of traditional show bugs inside the center: Madagascar Hissing Cockroaches, Australian Giant Cockroaches, Asian Walking Sticks, Malaysian and Asian Millipedes, and Emperor Scorpions. Box Elder Bugs are always on display, which makes sense since the zoo is surrounded by Box Elder trees. There's a nice collection of furry tarantulas too: Orange Knee, Zebra, Chilean Rose Hair and a Curly, whose hair really curls.

In the warmer seasons the zoo adds invertebrates such as Sphinx Moth caterpillars and Tomato Horn Worm caterpillars, also known as Potato Worm caterpillars in these parts since farmers produce plenty of potatoes here.

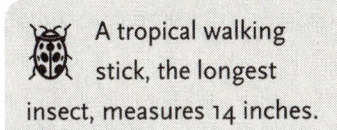

> A tropical walking stick, the longest insect, measures 14 inches.

Throughout the center there are microscopes and other scientific equipment. Enlarge your experience with magnified views of bugs, wings, hair and skins. Look for special programming in the summer and a day camp with a week themed with bugs.

Don't miss out on the rest of the live animal exhibits. The zoo features 600 animals from 125 species from around the world with native North American critters such as bison, Mountain Lions, moose and elk as well as exotic creatures such as yaks, aoudads and lemurs.

Hours: (Summer) 10 a.m. to 8 p.m. daily, late April to through October. (Winter) 1 p.m. to 5 p.m. Fri.–Sun. late October to late April, weather permitting.

Admission: (Summer) $4.50 adults, $2 ages 2–12, free for ages 2 and under. (Winter) $3.50 adults, $1.50 ages 2–12, free for ages under 2.

Entrance Barn

Red River Zoo

4220 21 Ave SW, Fargo, ND 58104
701-277-9240
www.redr.rzoo.org

Visit North Dakota after the winter thaw and you'll get a chance to tour this eight-acre wildlife habitat brimming with animals native to the Earth's northern climes. Meet Prairie Dogs, Chinese Red Pandas and Eurasian Wild Boars. Before you do, you'll have to walk through the zoo's entrance shaped like a big red barn and come face-to-face with some creepy critters.

The zoo is one of the few places in North Dakota where you can see live Indian Walking Sticks, Emperor Scorpions and Hermit Crabs. Look for the mock-up of a typical American kitchen and watch as hundreds of Madagascar Hissing Cockroaches scuttle along a white windowpane, a toaster and a half-eaten bagel. Compare these crawling creatures to their cousins, the South American Giant Cockroaches who climb from floor to ceiling in a 50-gallon aquarium. You'll find the Emperor Scorpions in aquariums sunk into counter tops and lit so you see them from both sides. The zoo also has a multitude of tarantulas of varying colors and sizes in a three-tiered exhibit. If your kids are in Fargo for the summer, sign them up for the **Zooniversity** camp where they get to compete in cockroach races.

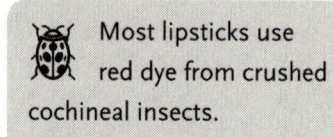

> Most lipsticks use red dye from crushed cochineal insects.

Hours: 10 a.m. to 8 p.m. daily, from first Saturday in May to last Sunday in September; 10 a.m. to 5 p.m. after Labor Day.

Admission: $5 adults, $4 seniors (60+), $3 ages 2–14, free for ages under 2.

Ohio

World of the Insect

Cincinnati Zoo & Botanical Garden

3400 Vine St., Cincinnati, OH 45220
513-281-4700 or 800-94-HIPPO
www.cincyzoo.org

When **World of the Insect** opened in 1978, it was only the second major arthropod exhibit in the U.S. and the first zoo to feature a major bug exhibit. Its maze of 66 displays housing millions of live bugs is still one of the largest and most diverse collections in the country.

Walk along a winding path to get close to the creepy crawlies from 10-inch-long walking sticks to pin-sized springtails. You'll see everything from beetles to butterflies to Bullet Ants, one of the fiercest ants in the world. Hailing from Central and South America, these giant ants pack a painful sting—30 times worse than a bee's. Collected by the zoo's entomologist extraordinaire Randy Morgan, the Bullet Ant colony was the first in the world at a zoo and the only one in the U.S. The colony of inch-long ants is divided between two glass cases; one gives a close-up view and the other simulates a rain forest where they forage for food.

Joining the unusual ants are Peruvian Fire Sticks, relatives of walking sticks that feed on the ferns of Peru. Marked by bright colors, the fire sticks secrete a protective chemical which smells and tastes like rubber cement. The colony earned the zoo a coveted Edward Bean Award from the American Zoo and Aquarium Association. The zoo also garnered the award four other times for being the first North American zoo to breed and rear in captivity the Goliath Beetle, Giant Southeast Asian Walking

Stick, the Hercules Beetle and the Harlequin Beetle.

There's more to see including a butterfly aviary, a Leaf Cutter Ant colony, and what may be the largest Giant Centipede in captivity. The foot-long arthropod has spent several years at the zoo.

Interactives are an integral part of the bug zoo, encouraging visitors to take a hands-on approach to learning about invertebrates. Step on the scale and discover what your weight is in bugs and pull out drawers to find the answers to buggy questions.

World of the Insect also features insectivores, animals that eat insects. Look for Tree Shrews, Green Snakes, Poison Arrow Frogs and several spiders. Also on display are the bizarre looking Naked Mole Rats, hairless rodents that live in underground colonies similar to termites.

Hours: 9 a.m. to 5 p.m. daily, Memorial Day to Labor Day; 9 a.m. to 4 p.m. Labor Day to Memorial Day. Closed Thanksgiving, Christmas Eve, Christmas Day, New Year's Eve, and New Year's Day.

Admission: (Main Zoo) $11.50 adults, $9 seniors, $6 ages 2–12, free for ages under 2. (Children's Zoo) $1 per person.

Parking: $6.50

The Crawl Space

Toledo Zoo

2700 Broadway, Toledo, OH 43609
419-385-5721
www.toledozoo.org

Head for the **Diversity of Life** section where insects invade **The Crawl Space**. About 25 species of invertebrates live in two rooms featuring a tank wall, terrariums with cutout views and freestanding displays that give you a 360-degree view of bugs.

There are Rhino Beetles from Poland, four main species of stick insects, neon green African Mantids, Cuban Green Cockroaches, and both Red Eyed and White Eyed Assassin Bugs. Orb Weaving Spiders from Madagascar reside in a big freestanding glass case so you can see their beautiful webs. Other creepy crawlies include Madagascar Hissing Cockroaches, Death's Head Cockroaches, Giant African Millipedes, and four tarantula species: Mexican Red Leg, King Baboon from Africa, Zebra from Costa Rica and the Giant Bird Eater.

The **Diversity of Life** area is also home to koalas, bats and Naked Mole Rats.

Hours: 10 a.m. to 5 p.m. daily, May 1 to Labor Day; 10 a.m. to 4 p.m. Labor Day to April 30. Closed Thanksgiving, Christmas and New Year's Day.

Admission: $8 adults, $5 seniors (60+) and ages 2–11, free for ages under 2.

Butterfly House

1145 Obee Rd., Whitehouse, OH 43571
419-877-2733
www.butterfly-house.com

When Duke Walters first opened up his magnificent house of butterflies it was only available to the public on weekends. But the site became so popular that it's now open daily from May 1 to September 30.

There's no question as to why it's a sought after place to visit. Duke built the house to showcase how beautiful gardens can attract blooms of butterflies. And he knows how to do it. Adjacent to his **Obee Rd. Garden Center**, which specializes in butterfly flowers and native flora, the butterfly house is a model of beauty. Red bricks form the 30' by 60' enclosure's walkway with flowering foliage spilling along the edges. Blooming planters hang from the ceiling. It's a serene setting with a waterfall bubbling and more than 500 butterflies from around the world fluttering around you. That's an incredible number of flying flowers flittering at once, particularly for a small non-institutional butterfly house.

What's even more exciting is the number of species you'll encounter. Since many species of butterflies only live two to three weeks as adults, a new species is introduced into the house every week. Throughout the season more than 50 species are released so return visits reap rewards of a new delicate creature to admire. During your self-guided tour you'll see several exotics: Ruddy Daggerwings, Mosaics and Isabellas from Central and South America, Banded Purple Wings and Malachites from Costa Rica and Orange Tigers from El Salvador. North American species include Tiger Swallowtails, Buckeyes, Cloudless Sulphurs, Sleepy Oranges, Red Admirals, Gulf Fritillaries, Zebra Longwings, Eastern Black Swallowtails and Spicebush Swallowtails. You can also watch for local butterflies in the gardens outside the enclosure.

Located in rural Ohio, the butterfly house makes visitors welcome. Picnic tables are available so you can make your visit an afternoon adventure. Be sure to visit the garden center to learn how to create your own backyard butterfly garden. Staff members are available in the butterfly house and the garden center to answer questions. Every Saturday during September there is a **Monarch Butterfly Release** at noon where locally raised Monarchs are tagged before migrating to Mexico.

Hours: 10 a.m. to 5 p.m. daily, 12–5 Sun., May 1 to September 30.

Admission: $5 adults, $4 seniors (65+) and ages 5–12, free for ages 4 and under when accompanied by an adult.

Blooms & Butterflies

Franklin Park Conservatory & Botanical Garden

1777 E. Broad St., Columbus, OH 43203
614-645-TREE or 800-214-PARK
www.fpconservatory.org

Imagine being surrounded by 10,000 free-flying South American and North American butterflies. You don't have to imagine it. You can do it when this venerable conservatory opens its indoor **Pacific Island Water Garden** to living colors beyond imagination. With its tropical environment and cascading foliage, the garden is perfect for this seasonal sensation.

More than 25 species spiral around fragrant Plumeria and Jasmine blossoms, tropical fruits and flowing waterfalls. Among them, iridescent Blue Morphos and their cousins, shimmering White Morphos. Behold Tiger Swallowtails, White Peacocks, Orange Barred Sulphurs, Orange Tigers,

Bluewings, Zebra Longwings, Doris Butterflies, Crackers and Julias.

Discover the wonders of metamorphosis in the conservatory's visible butterfly hatchery. Learn how to create a butterfly garden, take stunning photos of winged beauty and identify butterfly species when you sign up for educational classes.

Look for topiaries in the shape of butterflies throughout the botanical garden and enjoy the kaleidoscope of blooming bulbs including Tulips, Crocuses, Narcissus, Lilies and Lilacs in the conservatory and surrounding park.

Hours: 10 a.m. to 5 p.m. Tue.–Sun., 10 a.m. to 8 p.m. Wed., 10 a.m. to 5 p.m. certain holiday Mondays. Closed Thanksgiving, Christmas and New Year's Day.

Admission: $6.50 adults, $5 students and seniors, $3.50 ages 2–12, free for ages under 2.

The Magical World of Butterflies

Krohn Conservatory

Eden Park, 1501 Eden Park Dr., Cincinnati, OH 45202
513-421-5707
www.cinci-parks.org/parks/krohn/main.html

If you're visiting Cincinnati during early May to late June, be sure to take in the spectacular swirls of exotic colors at this annual free-flying butterfly event. For six weeks the conservatory showcases some of the world's most exotic species that weave a living quilt among 5,000 plants from around the world.

Located in an historic 186-acre park, the 1933 conservatory features hundreds of North American butterflies in the showroom while more than 30 tropical species from Central America, Malaysia and Africa zip through three separate gazebos. Look for the magnificent Birdwing Butterfly with its seven-inch wingspan.

Hours: 10 a.m. to 5 p.m. daily, early May to late June.

Admission: $5 adults, $4 seniors, $3 ages 5–17, free for ages 4 and under.

Butterfly House

Cox Arboretum

6733 Springboro Pike, Dayton, OH 45449
937-434-9005
www.metroparks.org

Summer is a superb time to visit this 170-acre botanical park because that's when the 2,100-square-foot native butterfly house opens. A living showcase of butterflies common to southwest Ohio, the screened house is open to the elements allowing the fluttering flowers and flowering flora to experience natural environmental changes.

All the butterflies inside are collected in the wild, making the house one of the few in the U.S. that displays butterflies this way. The setting permits the butterflies to continue their natural life cycle so you can observe eggs, caterpillars and chrysalides. When the butterfly house closes in the fall, all the live butterflies are released outdoors.

More than 40 butterfly species are common to this area, so you should see a good variety in the enclosure located near the wood's edge. Among them: American Coppers, Spring Azures, Orange Sulphurs, Dusky Wings, Northern Broken Dashes, Tiger Swallowtails, Great Spangled Fritillaries, Milbert's Tortoise Shells, Red Spotted Purples, Gray Hairstreaks, Little Wood Satyrs and Eastern Snout Butterflies.

Be sure to stroll the grounds featuring thematic gardens that highlight plants that thrive in southwestern Ohio, including a butterfly host garden and a newly designed butterfly nectar garden. There are also 1.5 miles of trails through woodlands and meadows with an observation deck that provides a scenic lookout.

Hours: (Butterfly House) 10 a.m. to 4 p.m. Tue.–Wed., Noon to 4 p.m. Sat.–Sun., by appointment 10 a.m. to 4 p.m. Thu. July 1 to early September. (Arboretum & Gardens) 8 a.m. to dusk. (Visitor's Center) 8:30 a.m. to 4:30 p.m. Mon.–Fri., 1 p.m. to 4 p.m. Sat.–Sun.

Admission: Free.

The Rainforest

Cleveland Metroparks Zoo

3900 Wildlife Way, Cleveland, OH 44109
216-661-6500
www.clemetzoo.com

Thunderclaps signal that you're near one of the largest rain forest exhibits in the U.S. Every 12 minutes a tropical rainstorm hits the two-acre exhibit complete with lightning, wind and rain. Across from the storm are displays of rain forest insects.

A Leaf Cutter Ant colony is the biggest display with a video camera that you maneuver to get the best view of the foraging and farming insects. The ants are magnified and shown on adjacent video screens. The huge glass terrarium reveals the ants gathering leaves and lets you peek inside their underground tunnels.

Located on the lower level of two, the ants are joined by other rain forest bugs: Dead Leaf Mantids, Madagascar Hissing Cockroaches, African Millipedes, walking sticks, centipedes, scorpions and tarantulas. Don't miss the 12-foot American Crocodile next to the insects, with split-level steps so you can go eye-to-eye with the big beast.

Be sure to visit the rest of the 165-acre zoo, which is home to the largest collection of primates in North America.

Hours: 10 a.m. to 5 p.m. daily, until 7 p.m. Sat.–Sun. from Memorial Day to Labor Day. Closed Christmas Day and New Year's Day.

Admission: $8 adults, $4 ages 2–11, free for ages under 2.

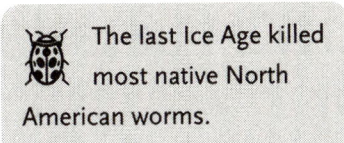

> The last Ice Age killed most native North American worms.

Eleanor Armstrong Smith Glasshouse

Cleveland Botanical Garden (closes November 2002; reopens Fall 2003)

11030 East Blvd., Cleveland, OH 44106
216-721-1600
www.cbgarden.org

These magnificent public gardens close in November 2002 so they can metamorphose into a vibrant new environment that includes a shimmering 18,000-square-foot glass conservatory that will house more than 50 species of butterflies. The butterflies are part of the contrasting environments that will represent two of the world's most fragile ecosystems when the gardens re-open in Fall 2003: the spiny desert of Madagascar and the cloud forest of Costa Rica.

The innovative biome design underneath the crystal peak will also include other bugs such as Leaf Cutter Ants plus birds, amphibians and reptiles. The glass house will immerse visitors in the sights, sounds and smells of these areas.

Blue Morphos and other exotic winged rainbows will be found in the Costa Rican environment which will feature a 40'-tall **Great Cloud Forest Tree**, also known as a Strangler Fig. Tucked inside the tree's hollow base will be three exhibits: a Whip Scorpion, a tarantula and an Emerald Swift Lizard.

The botanical gardens will remain open from April to October 31, 2002.

Oklahoma

Wings of Wonder

Tulsa Zoo

6421 East 36th St. North, Tulsa, OK 74115
918-669-6600
www.tulsazoo.org

Billed as a **Living Museum**, this zoo in the heart of Oklahoma features hundreds of North American butterflies in its 30' by 96' mesh-covered seasonal butterfly house. Look for Goatweed Leafwings, Hackberries, Great Purple Hairstreaks, Red Spotted Purples, Silver Spotted Skippers, Zebra Longwings, Question Marks, and Queens among the 30 species flying here. There are also two species of moths: Cecropias and Lunas.

As you walk through the landscaped exhibit try to spot the numerous Swallowtails: Black, Eastern Tiger, Giant, Palamedes, Pipevine, Two-Tailed and Zebra. Then look for the **Chrysalis House** displaying hundreds of butterfly pupae imported from butterfly farms.

While in the exhibit, discover the differences between butterflies and moths and find out what Oklahoma's 10 most common butterflies are. Learn about the butterfly lifecycle and how to attract these natural beauties to your own backyard. If you've got questions, there are zoo docents and staff to provide answers. Outside the enclosure, a garden full of butterfly-attracting plants entices local varieties to spend some time there.

On opening day each year the zoo celebrates with a rainbow of butterfly activities. One year the zoo encouraged guests to chalk butterflies on the sidewalks, handed out butterfly tattoos and presented a Native American performance relating to butterflies.

After boogying with the butterflies, head for the **Tropical American Rain Forest**, a living re-creation of Central and South American rain forests. A simulated **Kapok Tree** stretches 50 feet inside the 15,000-square-foot exhibit, which includes a host of invertebrates. Among the Black Howler Monkeys, Jaguars and Anacondas you'll find Australian Walking Sticks, Wood Roaches, Madagascar Hissing Cockroaches, a Leaf Cutter Ant colony, beetles, several scorpion species and tarantulas, millipedes, crickets and grasshoppers.

Hours: (Zoo) 9 a.m. to 5 p.m., daily. Closed Christmas Day and a selected day in June. (Butterfly House) mid-May to early October, weather permitting.

Admission: $6 adults, $4 seniors (55+), $2 ages 3–11, free for ages 2 and under.

Butterfly Garden

The Oklahoma City Zoo

2101 NE 50th St., Oklahoma City, OK 73111
405-424-3344
www.okczoo.com

Nature takes flight at Oklahoma's other big city zoo. Its 20,000-square-foot open-air butterfly garden is packed with 15,000 plants. Most of them are nectar producers that attract adult butterflies looking for nourishment. Look for Oklahoma's state butterfly, the Eastern Black Swallowtail on the Phlox and Milkweed while its caterpillar can be found on the Queen Anne's Lace and Rue.

Monarchs head for the Milkweed in the garden, while Viceroys, which mimic Monarchs, love the wildflowers. Look for Cabbage Whites hovering around the vegetables while Painted Ladies, which are native to the state's meadows, enjoy Red Clover, Zinnias and Shasta Daisies. Other butterflies you might see here: Coral Hairstreaks, Common Whites and Silvery Blues.

The garden's south corner features a tall grass prairie section planted on a hillside. The plants are the same as those found at the **Tall Grass Prairie Preserve** in northeastern Oklahoma. The state wildflower, the Indian Blanket (Gaillardia), is one of the flowering nectar sources for the butterflies. Other plants include Big and Little Blue Stem, Pitcher Sage, Prairie Coneflower, Verbena and Indian Grass.

Hours: 9 a.m. to 6 p.m. daily, Memorial Day to Labor Day; 9 a.m. to 5 p.m., winter.

Admission: $6 adults, $3 seniors (65+) and ages 3–11, free for ages under 3.

Oregon

Insect Zoo

Oregon Zoo

4001 SW Canyon Rd., Portland, OR 97221
503-226-1561
www.oregonzoo.org

There are more than 1,000 animals at this 60-acre zoo, and among them a variety of invertebrates that live in a small wooden house next to the **Penguin Habitat**. On the dark shady east side of the building are bugs that creep and crawl through the forest of the Pacific Northwest: domestic millipedes, Boring Beetles, Cave Crickets and Caddis Fly larvae. Sometimes the zoo sets up a local pond ecosystem and displays aquatic bugs. The front of the zoo, which gets direct sunlight, houses arthropods from arid climates: Mexican Red Knee Tarantulas, scorpions and centipedes. The third side of the zoo is home to tropical rainforest bugs: Giant African Millipedes, Madagascar Hissing Cockroaches and walking sticks.

During the summer the **Insect Zoo** patio is buzzing with activity.

Educational staff and volunteers are available to answer buggy questions, bring out critters for visitors to touch and help kids with a build-a-bug craft. A table is set up with microscopes and hand lenses for further bug exploration. Butterfly-attracting plants landscape the building and you get a good view of butterflies in an adjoining flight cage.

Hours: 9 a.m. to 4 p.m. daily October 1 to March 31; 9 a.m. to 6 p.m. April 1 to September 30. Closed Christmas.

Admission: $7.50 adults, $6 seniors (65+), $4 children ages 3–11, free for ages 2 and under. Free the second Tuesday of every month from 1 p.m. until closing.

Butterflies Forever

Under the Bridge on the Astoria Riverfront

260 Bay St., Astoria, Oregon 97103
www.bforever.org

Visit the only butterfly house in Oregon, where 300 free-flying North American butterflies flutter in a netted garden. Originally opened in Seaside, **Butterflies Forever** has moved its seasonal enclosure to **Under the Bridge on the Astoria Riverfront**, an area undergoing redevelopment. The move in 2002 year to a spot near the Columbia River is the beginning of a new life for the butterfly house, which hopes to metamorphose into a permanent glass pavilion where 1,000 tropical butterflies will soar.

The 6,000 to 8,000-square-foot glass conservatory will be a first for the Northwest. The butterfly pavilion will be only part of a 20,000-square-foot complex that will include educational classrooms, a theater, display areas and a store. The non-profit **Butterflies Forever** welcomes donations to help get closer to building a domed tropical garden full of living rainbows.

Meanwhile, enjoy the sights of local winged wonders and learn about the threatened Oregon Silverspot Butterfly and how you can help preserve the species. Some local businesses and civic groups sponsor community days at the butterfly house by offering free or reduced admission, so be sure to ask how you can participate.

Hours: 10 a.m. to 5 p.m. daily, Memorial Weekend to September 30.

Admission: $4 adults, $2.50 seniors (60+), $1.50 ages 4–12, free for ages 3 and under with a paid adult.

Bugs Alive, A Living Arthropod Exhibit

Cordley Hall
Oregon State University

Department of Entomology, Corvallis, OR 97331
541-737-5520
roycel@bcc.orst.edu
www.osu.orst.edu

Did you know that Oregon is home to scorpions? The native Wood Scorpion is common throughout the Willamette Valley, according to Lynn Royce, an insect identification specialist who heads the university's bug zoo. Several other arthropods join the scorpion in a small room that opened up for the bug business just a few years ago: walking sticks, spiders, beetles, termites, ants, silverfish, cockroaches, millipedes and centipedes.

The zoo's big public access comes twice a year in the spring and fall during the university's free **Museum Days** event (see Bugged Out chapter). Since the zoo primarily serves as a teaching tool and as a traveling educational exhibit to local schools, it does not have any set hours. However, if you contact Lynn, you may be able to set up an appointment to view the critters, even touch a few of them. Keep in mind that the room is designed for only 3 to 4 people to visit at a time and that the zoo runs on donations and grants.

Hours: During regular university hours.
Admission: Donation.

> Spider silk is the strongest natural fiber.

Pennsylvania

Butterflies

The Academy Of Natural Sciences

1900 Benjamin Franklin Parkway, Philadelphia, PA 19103
215-299-1000
www.acnatsci.org

Come in from the chill of Philadelphia's winter and discover the splendor of exotic colors at the region's only indoor butterfly house featuring tropical flying flowers from Costa Rica, Malaysia and Africa. Raised on rain forest butterfly farms, these winged wonders represent an alternative to clear-cut agriculture in areas already hurting from deforestation. The exhibit highlights not only the beauty of these exotic species but butterfly conservation as well. The museum recently added butterflies from a farm project in Kenya's Arabuko/Sokoka Forest as a way to show that the African forest can utilize its wild land without depleting it. While you wait in line to enter the house, read the over-sized postcards sent from the rain forest's human inhabitants and watch a video about the benefits of butterfly farming. Inside, gliding gems flutter freely as you walk along a path complemented by a stream and blooming Hibiscus plants. About 50 to 60 butterflies are out at any one time. School groups can take advantage of the **Backyard Butterfly Lab** that encourages students to create wildlife habitats in their urban communities.

Hours: 10 a.m. to 4:30 p.m., Mon.–Fri.; 10 a.m. to 5 p.m. Sat.–Sun, holidays. Closed Thanksgiving, Christmas, and New Year's Day.

Admission: $9 adults, $8.25 seniors, $8 ages 3–12, includes museum admission.

Butterfly House

Hershey Gardens

170 Hotel Rd., Hershey, PA 17033
717-534-3492
www.hersheygardens.org

The butterflies are sweet—but not made of chocolate—at these famous gardens named for Milton S. Hershey and his chocolate company. About 300 shimmering wings fill the 30' by 60' mesh enclosure on any given day, but you should call ahead to make sure the butterflies are flying before you visit.

Among the 25 species of North American wonders winging their way through Mexican Sunflowers, Purple Coneflowers and Spike Gayfeathers are Great Spangled Fritillaries, Eastern Tiger Swallowtails and Silver Spotted Skippers. Look for Malachites, Red Admirals and Julias among the nectar and host plants such as Butterfly Bush, Brazilian Verbena and Mealycup Sage.

Want to identify the nectar and host plants like Butterfly Weed and Egyptian Star-Clusters, or tell the difference between a Queen and a Monarch Butterfly? Volunteer "flight attendants" are available to answer questions. You can also refer to the butterfly and plant identification guides provided inside the house or buy them at the **Gardens Gift Shop**.

Observe the entire butterfly life cycle within the house as the delicately winged insects lay eggs, the caterpillars munch on host plants and the chrysalides hang waiting for the moment that their captured treasures break free.

Visitors are allowed entry on a first-come basis and are admitted by the House Attendant.

Constructed in 1998, the butterfly house was built using a wing from the original greenhouse built for M.S. Hershey in the 1930s. The 23-acre gardens opened in 1937 as a smaller Rose garden. That 3.5-acre historic garden still exists with nearly 275 varieties and 7,000 Rose plants.

In the near future, look for a themed **Children's Garden** that features a

Caterpillar Tunnel and Butterfly Garden. The fuzzy caterpillar shape will be cultivated from vines and will lead into the body of a butterfly with raised beds shaped like wings that are planted with butterfly-attracting flowers.

Hours: (Gardens) 9 a.m. to 6 p.m. daily March 31 through September, until 8 p.m. Fri.–Sun. Memorial Day to Labor Day, 9 a.m. to 5 p.m. daily October 1 to October 31. Closed November 1 to March 29. (Butterfly House) 9 a.m. to 6 p.m. daily, second Saturday in June through the third Saturday in September, weather permitting.

Admission: $6 adults, $5.50 seniors (62+), $3 ages 3–15, free for ages under 3.

The Insectarium

Steve's Bug Off Exterminating Company

8046 Frankford Ave., Philadelphia, PA 19136
215-338-3000
www.insectarium.com

What started off as the "catch of the day," in the front window of the pest control company has turned into 6,500-square-feet of the buggiest museum in the tri-state area. Owner Steve Kanya opened the doors to what was then an unusual museum in January 1992 and it has grown into a two-floor showcase of live bugs, preserved specimens and an outdoor learning garden.

See the kitchen and bathroom crawling with American Cockroaches, gaze at the tank filled with glow-in-the-dark scorpions and observe two working colonies: Honey Bees and live termites. There are plenty of other characters creeping inside terrariums: Madagascar Hissing Cockroaches, Mexican Red Leg Tarantulas, Emperor Scorpions, Goliath Beetles, Indian Walking Sticks, Praying Mantids, African Millipedes, Thorny Devils, Human Face Stink Bugs, Waterbugs, Velvet Ants, Camel Crickets and centipedes.

Sign up for a themed, guided tour for a complete experience. The tours include an age-appropriate movie, bug handling and gourmet insect tasting. Where else can you pet a Madagascar Hissing Cockroach, watch a tarantula slurp a cricket then chow down on a crispy mealworm yourself?

The only bug museum in the Philadelphia area is so popular that it holds two-hour buggy birthday parties on Saturday, buggy craft days and special monthly programs such as "Bug Babies" and "Buzzing About the Boardwalk." The museum also offers afternoon study programs and "Bug Scouts!" an after-school program with crafts, discussion and hands-on learning.

Hours: 10 a.m. to 4 p.m. Mon.–Sat.

Admission: $5 per person.

Butterfly Forest

Phipps Conservatory & Botanical Gardens

One Schenley Park, Pittsburgh, PA 15213

412-622-6914

www.phipps.conservatory.org

"Something's always blooming at Phipps," and during the spring and summer brilliant butterflies blossom at the only butterfly exhibit in western Pennsylvania. For five months, hundreds of native species spiral through the conservatory's **Stove Room**: Julias, Monarchs, Queens, Buckeyes, Red Admirals, Orange Sulphurs, Giant Swallowtails, Zebra Longwings and Gulf Fritillaries. The exhibit is planted with Passion Vine, Pentas, Jungle Geranium, Mexican Heather, Hibiscus and Coral Plants; sugar water and tropical rotting fruit provide additional nourishment.

Every week 325 butterfly chrysalides come in from four Florida butterfly farms. You can see them on display boards in the forest's hatching cases.

What's unusual about this butterfly habitat is that it happens inside a 13-room Victorian glasshouse that's 109-years-old. When it opened in 1893, the crystal palace of industrialist Henry Phipps was the largest conservatory in the U.S. To create an environment where the butterflies will thrive, the conservatory staff has had to be creative. For humidity in the forest, Vermiculite, a soil-less substance that retains water, is used in the bottom of the hatching cases. Air currents at the entrance and exit are used to keep the butterflies from bailing.

Besides butterflies, there are a host of other plants and gardens including a miniature Orchid collection, a **Discovery Garden** and **Japanese Courtyard Garden**. The gardens hold several family-friendly events including **DinoQuest** and the **Discovery Garden's Children's Festival**.

Hours: 9 a.m. to 5 p.m. Tue.–Thu. and Sat.–Sun., 9 a.m. to 9 p.m. Fri. Closed Mondays, Thanksgiving and Christmas.

Admission: $6 adults, $5 seniors, $4 students with I.D., $3 ages 2–12, free for ages under 2.

Rhode Island

Newport Butterfly Zoo

Middletown Butterfly Farm

1038 Aquidneck Ave., Middletown, RI 02842
http://community-2.webtv.net/butterflyzoo/doc2/

This may be the only farm in New England that rears, releases and sells butterflies. Its 30' by 100' screened greenhouse shaped like a Gothic arch features butterflies from around the world. Fluttering in the seasonal yet permanent butterfly house are White Peacocks, Giant Swallowtails, Paradise Birdwings, Julias, Zebra Longwings, Red Admirals, Viceroys and Variegated Fritillaries. Between 500 and 1,000 butterflies zip through the wildflower meadow interior landscape including Spicebush and Anise Swallowtails, Gulf Fritillaries, Monarchs, Queens, Painted Ladies and Postman Butterflies. There's also a pond with Butterfly Koi and a trained turtle that obeys a few simple commands.

Biology teacher and linguist Marc W. Schenck owns the farm and zoo and can give guided tours in French, Spanish and Russian. If it's a rainy day, count on the butterfly house to be closed since the butterflies hide in the plants on those days. Families, please note, the zoo has no restrooms.

Hours: 11 a.m. to 4 p.m. Tue.–Sun. mid-May to mid-September.

Admission: $4 adults, $2 ages 3–12, free for ages under 3. Add $1 to admission on weekends.

South Carolina

Butterfly Pavilion

Broadway at the Beach

1185 Celebrity Circle, Myrtle Beach, SC 29577
843-839-4444
www.butterfly-pavilion.com

Majestically jutting out over the strand, this phenomenal glass conservatory is as stunning as the live butterflies and birds inside. Located in a popular entertainment area of this coastal tourist town, the 22,500-square-foot pavilion is the only glass-enclosed building in the U.S. that includes a butterfly house featuring native and exotic butterflies, a discovery center with insects, reptiles and amphibians, a Lorikeet aviary, gift shop and restaurant. It may hold the largest collection of live land invertebrates in the country.

In the **Butterfly Conservatory**, 9,000 square feet is devoted to a lushly landscaped tropical paradise where more than 40 species of delicate delights dance among the enticing greenery. Watch as 2,000 butterflies and moths including Mother of Pearls, Emerald Swallowtails and Blue Spotted Emperors sip nectar and Viceroys, Julias and Banded Oranges bask in the spilling sunshine. Winding pathways offer several vantage points to view the Rice Papers, Buckeyes and Sailors. Sharing the towering 40-foot high space with the Archdukes, Blue Glassy Tigers and Spicebush Swallowtails are exotic birds: Kookaburras, Macaws and Pharaoh Quail and "bird of preys" such as Screech Owls and Red-Tail Hawks. You'll find more birds in the **Lorikeet Aviary** where playful rainbow colored parrots eat fruit nectar from a cup in your hand.

Head to the **Nature Zone Discovery Center** where you step into re-created cave crevices and wind up face-to-face with New Guinea Thorny Devil Sticks, Asian Centipedes, Death's Head Cockroaches, Giant Prickly Walking Sticks, Giant African Millipedes and Giant Cockroaches. Make your way to the **Explorer's Camp** where beautiful Cuban Green Banana Cockroaches reside with their cousins the Madagascar Hissing Cockroaches plus Pink Toe and Chilean Rose Tarantulas. You'll find more spiders in the **Arachnid Exhibit** featuring a Malaysian Gold Huntsman Spider, Mexican Beauty Tarantula, Goliath Bird Eater, Wolf Spider, Black Widow and a lone Emperor Scorpion. At the **Think Tank** discover some more arthropods: Giant Pond Snails, Velvet Ants and a Mexican Red Knee Tarantula.

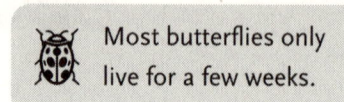

> Most butterflies only live for a few weeks.

Crawl into the terrestrial tunnel of South American Leaf Cutter Ants and watch as they parade giant bits of leaves, deposit them at their farm and turn them into fungus for food. You can also watch North American Honey Bees make their own golden nectar in an observation hive. For an extra dose of fun slip inside the re-creation of a giant tree and peer at poisonous tree frogs, toads and bullfrogs.

Hours: 10 a.m. to 4:30 pm. Mon.–Sat. 11 a.m.–4:30 p.m. Sat. Hours change during summer.

Admission: $9.99 adults, $8.99 seniors (55+), $7.99 ages 4–12, free for age 3 and under.

Butterfly House

Cypress Gardens

3030 Cypress Gardens Rd., Moncks Corner, SC 29461
843-553-0515
www.cypressgardens.org

Prepare for a more tranquil experience in this 170-acre park about 24 miles north of Charleston. Known as the Gem of Berkeley County, the county park promotes the beauty of a cypress swamp. Formerly part of **Dean Hall Plantation**, the area was used to raise rice and the old rice reservoir is now called the **Cypress Gardens Swamp**.

Opened in 1997, this was the first butterfly house in South Carolina. The 2,500-square-foot covered greenhouse style pavilion features 8 to 12 native species of butterflies at any one time. During peak season about 140 butterflies flutter around you in this intimate garden setting. Most of the time you'll see Malachites, Julias, Queens, Buckeyes, Monarchs, Sleepy Oranges, Cloudless Sulphurs, Zebra Longwings and Gulf Fritillaries. Swallowtails include the Black, Giant and Palamedes.

Flowering flora is abundant for both the adult butterflies and their caterpillars. There are Pentas, Lantanas, Butterfly Bushes and Zinnias for adult butterflies and Parsley, Mexican Milkweed, Passiflora and Snapdragons for the larva. You can watch the whole butterfly lifecycle if you look carefully for eggs, caterpillars and chrysalides. Some chrysalides are hung on a display board, too. Sit for a while on the benches and watch the delicate wings of both butterflies and Purple Honeycreepers—nectar-feeding birds from South America. There's also a small pond with Koi fish, turtles and ducks.

Look for the small Honey Bee observation hive with the bees going through a tube in a wall to collect pollen and bring it back to the hive. There are more bugs in an arthropod exhibit that holds a live Emperor Scorpion, Giant American Cockroaches and crayfish. Look for preserved invertebrate specimens of Giant Peruvian Centipedes, Giant African Millipedes and Horseshoe Crabs.

After your butterfly stay be sure to wander the garden paths and nature trails of the gardens. In spring there are thousands of brilliantly blooming Azaleas, Daffodils and Atamasco Lilies. Loop around the black water lakes to view the **Wedding Garden**, **Camellia Garden**, **Garden of Memories** and **Woodland Gardens**. Float through the waters on bateaus (flat bottom boats) and say hi to an alligator or two. Visit the aquarium and reptile center or spend your time looking for the teeming wildlife: herons, egrets, woodpeckers, warblers, turtles and otters.

Hours: 9 a.m. to 5 p.m.; no admittance after 4 p.m.

Admission: $9 adults, $8 seniors (65+), $3 ages 6–12, free for ages 5 and under. One price for all exhibits and activities including boat rides.

Butterfly Kingdom (opening 2003)

430 William Hilton Pkwy., Pine Station, Suite 511, Hilton Head, SC 29910
843-671-3200
www.butterflykingdom.com

Waiting in the wings is a planned 84,000-square facility that's promising to be the largest butterfly conservatory in North America. Located at the gateway to Hilton Head Island, this butterfly house is anticipated to open in 2003. The developers of this enormous project have big plans. Not only will this feature the butterfly conservatory but the site will be a natural science education center with an extensive insect zoo, nocturnal moth exhibit, an insect discovery and learning center, a university level research lab, a 400-seat 3D giant screen theater, gift shop and restaurant.

The 75-foot-tall glass greenhouse is designed to look like a butterfly. A botanical garden overflowing with exotic plants, it's expected to feature more than 120 species of butterflies in a 13,000-square foot conservatory that will also house butterfly, moth and insect displays. The plans include a 3,500-square-foot learning center, a museum with more than 100,000 mounted insect specimens and an 80-seat multimedia theater.

About 15 acres of outside gardens will surround this environmental learning facility. Flowers and plants from five continents will be represented along with fountains, educational gardens, lakes, picnic areas and playgrounds. Future plans include an entire resort with a water and surf park, golf driving range, hotels, restaurants and shopping. Until 2003, visit the conservatory's nature store, the **Butterfly Kingdom Emporium**, at Pineland Station Mall, Suite 201, William Hilton Parkway. Call 843-689-6500.

South Dakota

Sioux Falls Butterfly Garden

Outdoor Campus

Sertoma Park, 4500 S. Oxbow Ave., Sioux Falls, SD 57106
605-362-2777
www.outdoorcampus.org

Walk into 6,000 square feet of butterfly buffet when you enter this nature museum's bountiful butterfly garden. The South Dakota Federated Garden Clubs planted this home for native winged wonders by using a landscaping recipe that provides a full menu of butterfly-attracting flora. Question Marks, Commas, Black Swallowtails, Viceroys, Monarchs, Painted Ladies, Melissa Blues, Tiger Swallowtails, Variegated Fritillaries, Clouded Sulphurs, Hackberry and Alfalfa Butterflies are some of South Dakota's 171 species that dine on the bouquet of nectar-bearing plants. Look for Monarda, Pentas, Daisies, Verbena, Cleome, Coneflowers, Black-Eyed Susans and Phlox in the garden. There's also a separate section planted with host plants such as Parsley, Prickly Ash, Clovers, Lupines and Violets for the butterflies' caterpillars.

During summer months, butterflies can be found inside the museum as well. Monarchs, Painted Ladies and Black Swallowtails are raised so visitors can see all the life stages. They're released once a day and any visitors who are there can help with the butterfly release. You may also see a variety of other bug life, since visitors bring all sorts of critters in jars, boxes and shoeboxes to show or to get identified. You never know what will be there for the day. All critters are returned to the outdoors.

The museum participates in **Monarch Watch** and tags butterflies for release. If you see Monarchs frequenting your Sioux Falls yard, the **Outdoor Campus** will bring a volunteer team out to tag or teach you and your family how to do it.

Be sure to visit on **Butterfly Day**, when the campus soars with extra activities. Take guided tours of the butterfly garden, learn butterfly gardening tips, watch a butterfly puppet show, see great displays of butterfly specimens and listen to nature songs.

Hours: Garden—8 a.m. to 10 p.m. daily. Museum—8 a.m. to 5 p.m. Mon.–Fri., 8 a.m. to 4 p.m. Sat., 1 p.m. to 4 p.m. Sun.

Admission: Free.

Sertoma Club Butterfly House
(opening 2002)

Sertoma Park

West 49th St. at Big Sioux River, Sioux Falls, SD 57101
605-368-5447 or 605-335-5790
www.sfnoonsertoma.com/butterfly.html

New neighbors are moving next door to the **Outdoor Campus** and they're flying in to set up residence. The tropical flyers should be decorating their new home by Fall 2002, according to the Noon Sertoma Club that's responsible for constructing the largest butterfly house in the Great Plains region. The butterflies will live in an 8,000-square-foot glass-roofed conservatory that's part of a larger 18,000-square-foot facility.

A civic club, Sertoma (which stands for Service To Mankind) is building the butterfly center in its 180-acre **Sertoma Park**. Located in an old oxbow of the Sioux River, the butterfly house is also supported by the

South Dakota Game, Fish and Parks Department and the City of Sioux Falls. The park includes the **Outdoor Campus**, play areas and picnic shelters. The year-round butterfly house will be stocked with hundreds of exotic tropical butterflies and will include educational butterfly displays, classrooms, a store and an eatery. Future plans call for an 80' by 90' insectarium called the **Crawl-a-seum**.

You can help contribute to the butterfly fund by contacting the Sertoma Club.

Nature Works

Kirby Science Discovery Center

Washington Pavilion of Arts and Sciences
301 N. Main Ave., Sioux Falls, SD 57104
605-367-7397
www.washingtonpavilion.org

Listen closely and you'll hear the buzzing of bees coming from the science center's **Nature Works** exhibit. The honey hive features text panels explaining bee behavior including their funky dances. Bees actually perform choreographed body movements like the "Waggle Dance" to give directions to the nearest flowerbed. Watch the bees come and go through a hollow tube, which runs through a window so they can collect pollen. Look for the puffy orange sacks around their back legs—these are their sacks of "gold" filled with pollen.

Located in the **Animals ALIVE** section of the exhibit, the bees aren't the only native creatures you'll encounter. There's also a tarantula, a Death's Head Cockroach colony, toads, turtles, skink lizards and a Hog-Nosed Snake named Apollo. Staff and volunteers will bring out the safe beasts for visitors to see up close and even handle. Just ask!

In the **Animal Senses** section, experience bug and bat vision with interactive exhibits. Visit the **Animal Sounds** area to match the tracks of bison, pheasants, coyotes, elks, deer, and eagles with the sounds the animals make.

Hours: 10 a.m. to 5 p.m. Tue.–Thu. and Sat., 10 a.m. to 8 p.m. Fri., Noon to 5 p.m. Sun. Closed most Mondays between Labor Day and Easter except for school holidays.

Admission: $6 adults, $5 seniors (62+), $4 ages 3 to 12, free for ages under 3.

Tennessee

Bees

Creative Discovery Museum

321 Chestnut St., Chattanooga, TN 37402
423-756-2738
www.cdmfun.org

Watch the bees waggle and wriggle on the second floor of the area's only museum to combine music, art, science and technology under one roof. Most observation Honey Bee hives are fairly standard in concept, but this one is significantly different. A traditional hive is actually located outside the museum on its roof, but one wall of the hive can be observed through glass in the **Optics Tower** area.

Master Beekeeper Mike Studer designed the indoor portion of the hive. It resembles a regular working hive in height with four levels for guest observation. The hive is worked from the back with each level pulled out like a dresser drawer. One frame of bees can even be removed for outreach programs at local schools.

The seven to 10,000 bees, a mix of Italian and New World Carniolan colonies, buzz to and from the museum from the second floor. See if you can pick out the yellow and black Italian Bees, which build lighter combs and gather more pollen and the gray to black bodied Carniolans from the Czech Republic, which build less burr comb and produce fewer drones.

Three young beekeepers (kids ages 12 and up) and three staff members are able to care for the hive because they took a Master Beekeeping program given at the museum by Dr. John Skinner from the University of Tennessee who is also the State Apiarist. Beekeeping is a dying art and the museum hopes to promote it by offering the Master Beekeeper certification program.

Be sure to visit during the annual **Honey Harvest Festival** in July when national and local beekeepers and suppliers gather at the museum. Taste a variety of honeys, watch honey be harvested, roll a beeswax candle, build a beehive and the play the bee game.

Hours: 10 a.m. to 5 p.m. Tue. and Thu.–Sat., Noon to 5 p.m. Sun., closed Wed., September through February. 10 a.m. to 5 p.m. Mon.–Sat., Noon to 5 p.m. Sun., March through May; 10 a.m. to 6 p.m. daily, Memorial Day through Labor Day.

Admission: $7.95 adults, $4.95 ages 2–12, free for ages under 2.

Invertebrate Inn

Ijams Nature Center

2915 Island Home Ave., Knoxville TN 37901
865-577-4717
www.ijams.org

A former butterfly habitat, this cloth covered greenhouse is now a self-interpretive, natural outdoor bug house. Located on a hill at this 80-acre park in East Tennessee, the house is home to native plants and invertebrates. The inside garden is designed to show visitors what they can do in their own backyard with plants such as Lantana, Impatiens, Butterfly Weed, Butterfly Bush and Sweet Shrub. An herb area has butterfly food plants like Dill and Fennel. Walk on the paver stones in the garden to look for local bugs such as grasshoppers, crickets, Praying Mantids and walking sticks. About five different species of ladybugs can be found here. See if you can spot one of the black and yellow Writing Spiders known for weaving zigzags in their large silks and made famous in the book, *Charlotte's Web*. On the left hand side of the exhibit is a game with doors you open to discover general invertebrate information. **The Inn** is located on the nature center's paved universal trail and is hopping with activity in late July and August.

Hours: (Gardens) 8 a.m. to Dusk, daily. (Nature Center) 9 a.m. to 4 p.m. Mon.–Fri., Noon to 4 p.m. Sat., 1 p.m. to 5 p.m. Sun.

Admission: Free.

Bugs!

Knoxville Zoo

3500 Knoxville Zoo Dr., Knoxville, TN 37914
865-637-5331
www.knoxville-zoo.org

Did you know that a ladybug could play dead? Or that a centipede distracts predators by popping off a wriggling leg? Find out about these and other fascinating facts when the **Knoxville Zoo** gets infested by **Bugs!** A temporary exhibit, **Bugs!** summer home is inside the zoo's **Pilot Traveling Exhibit Center** and features about 30 species of arthropods. You'll meet Giant Water Bugs known as "toe-biters," Assassin Bugs, centipedes, stick insects, scorpions and tarantulas. The exhibit will be surrounded by a variety of mammals, birds, reptiles and even plants that love to snack on insects. After the summer, **Bugs!** will travel to another zoo near you.

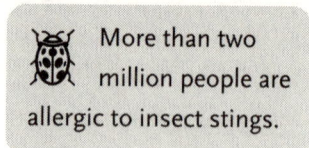

> More than two million people are allergic to insect stings.

Hours: 10 a.m. to 4:30 p.m. daily, closed Christmas.

Admission: $7.95 adults, $5.95 seniors 62+, $4.95 ages 3–12, free for ages 3 and under.

Parking: $3.

Backyard Wildlife Center

Lichterman Nature Center

Pink Palace Family of Museums, 5992 Quince Rd., Memphis, TN
901-767-7322
www.memphismuseums.org/nature.htm

Experience a whole new way to go buggy at this 65-acre environmental oasis in the center of urban Memphis. Early in 2002 the center hopes to open its **Backyard Wildlife Center** complete with a 4,400-square-foot *netted* re-creation of a mid-South meadow. Even though the **Living Meadow Habitat** is "under wraps" the synthetic roof keeps it open to the elements and will be stocked with a variety of insects, including butterflies.

A handicapped-accessible concrete trail lets you get close to the native plants that provide food and shelter for the butterflies, insects and arachnids that call this home. **Reader Rails**, (low-mounted graphic panels), located along the trail let you identify the bugs and plants you encounter. Information on the rails will change as the bug and plant population do.

You'll also find an observation Honey Bee hive along the trail. Bees buzz in and out of the netting through a tube and a **Reader Rail** explains bee biology from life cycles to nectar gathering to honey production. Learn about pollination, how the bees populate their hive and the dances they do to communicate with each other.

Inside the nature center you'll find live microscopic creatures from the **Litcherman Lake** as well as dragonfly larvae, crawfish, aquatic insects such as Water Beetles, Praying Mantids, walking sticks, venomous and non-venomous spiders, various beetles, scorpions and ants. Wendt Scopes let visitor get magnified view of bugs and Terrascopes use video cameras to get candid shots of bug activities.

> Daddy-long-legs are not spiders, but harvestmen.

Be sure to visit the **Living Lake Habitat** with cutaway underwater viewing of species living there and the interactive exhibits in the **Visitor's**

Center that focus on metamorphosis, seasonal changes, and the local flora and fauna.

Hours: 9 a.m. to 4 p.m. Mon.–Thu., 9 a.m. to 5 p.m. Fri.–Sat., Noon to 5 p.m. Sun.

Admission: $6 Adults, $5.50 seniors (60+), $4.50 ages 3–12, free for ages under 3.

Texas

Rainforest Pyramid

Moody Gardens

One Hope Blvd., Galveston Island, TX 77554
800-582-4673
www.moodygardens.com

Explore the wonders of Asian, African and American rain forests in a stunning 10-story glass pyramid that gleams on this Gulf Coast island. The steamy one-acre rain forest under glass is a living showcase of exotic plants and animals, just one of the spectacular environmental and scientific attractions in this 242-acre sub-tropical garden.

Living within and around the 55-foot-tall canopy that's almost as dense as a real rain forest, are 175 species of butterflies, birds, bats and fish. The animals and visitors relish the 2,000 types of flora that flourish in this humid environment such as African Violets, Cinnamon and Rosy Periwinkle. Fluttering among the flowering tropical plants are 1,000 to 2,000 elegant butterflies that take flight beneath the shimmering glass.

Watch as sparkling Blue Morphos, big Owl Eyes and Zebra Longwings emerge from their glittering chrysalides in the **Butterfly Hatching Hut** then take a peek at the peculiar flying mammals in the **Bat Cave**.

Bugs also play a role in keeping the rain forest pyramid free of pests.

Look for ladybugs, ladybug larvae and tiny parasitic wasps that provide a pesticide-free way of controlling wayward insects.

Moody Gardens is a terrific family destination complete with its own hotel. The gardens feature America's first **IMAX 3D Theater**, a paddle-wheel boat and a sandy beach. Both its aquarium—one of the world's largest—and the **Discovery Museum** are housed in smaller pyramids.

Hours: 10 a.m. to 6 p.m. Sun.–Thu., 10 a.m. to 8 p.m. Fri.- Sat.

Admission: $7.95 adults, $6.75 seniors (65+), $5.75 ages 4–12, free for ages 3 and under. $28.95 for special one-day passes that include the Rainforest, IMAX 3D, IMAX Ridefilm, Aquarium, Paddlewheeler and Palm Beach (seasonal).

Cockrell Butterfly Center

Houston Museum of Natural Science

One Hermann Circle Dr., Houston, TX 77030
713-639-4601
www.hmns.org

Texas, being a big state, has not just one but *two* extraordinary glass butterfly houses. The **Cockrell Butterfly Center**'s shape looks like a truncated cone, 70 feet tall at its highest point and 105 feet wide at the base. Each of the 588 panes of glass that form the walls and ceiling were custom cut and none is a perfect rectangle.

What's inside for visitors is just as astounding. To reach the opening where 1,500 winged jewels fly, you enter through a cave-like opening studded with stalactites and stalagmites. You pass behind a transparent curtain of water that cascades 40 feet from an overhanging cliff. Streams of light shine on the steamy Central American rain forest setting as you walk beneath a canopy of tropical plants. As if parting vines in a real rain forest, you step through the lush vegetation on a path that leads to a clearing overflowing with flowering shrubs and brilliant blooms. And you've just reached the point where colorful clouds of exotic butterflies appear!

The center has permits to release up to 150 tropical, North American and native Texas species into the re-created rain forest, but only about 50 of them are frequent flyers. Look for Blue Waves, Blue Clippers and Blue Morphos; Red, Blue and Grey Crackers; Tawny Owls, Giants Owls and Purple Owlets; Shoemakers, Cattlehearts and Malachites; Gold Rims, Mormons and Mimes. About 30 percent of the beautiful bounty is reared at the museum while the rest come from butterfly farms in Central and South America and Asia.

As you leave the main level up the stairway that follows the cliff behind the waterfall, take a closer look at the wall. There are three openings encircled by Mayan carvings and inside you'll see cases with jewel-like chrysalides. On a good day you can observe newly emerged butterflies stretching their damp wings.

Exit into the 3,000-square-foot **Brown Hall of Entomology**, which showcases a portion of the museum's 100,000 preserved insects. Examine thousands of tropical butterflies, moths and beetles and explore several educational exhibits before visiting the rest of the museum.

Your buggy expedition doesn't end yet though. In fact, it just begins because before the entrance to the amazing butterfly conservatory you have to mingle with live creepers in the museum's **Insect Zoo**. Look for the delicate Orchid Mantid that mimics an Orchid bloom and the frightening Goliath Beetle with its three-horned carapace. Watch as the large Leaf Cutter Ant colony farms fungus for food and observe Madagascar Hissing Cockroaches, Giant Millipedes, centipedes and Vinegaroons. Texas natives include the state's three most dangerous arthropods: Black Widows, Brown Recluse Spiders and scorpions. Other natives on display: Lubber Grasshoppers, Ant Lions and a colorful colony of Velvet Ants also known as "cow killers" because of their nasty sting.

Hours: 9 a.m. to 6 p.m. Mon.–Sat., 11 a.m. to 6 p.m. Sun.

Admission: (Butterfly Center) $4 adults, $3 seniors and ages 3–11, $2.50 members. (Museum) $5 adults, $3 seniors and ages 3–11.

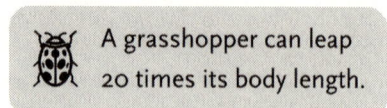

> A grasshopper can leap 20 times its body length.

Monarch Sanctuary

Abilene Zoo and Discovery Center

2070 Zoo Lane, Nelson Park, Abilene, Texas 79602
915-673-WILD
www.abilenetx.com/zoo/zoo_home.htm

A petite zoo with 800 animals representing 200 species, the primary focus of its 13 acres is on animals from the Southwestern United States, Central America, Africa and Madagascar. The zoo is also keenly aware of its own roots—plants native to South Texas are abundant and surround the exhibits. Two planted areas are designated as butterfly gardens and since they are purposely packed with juicy nectar-bearing plants that attract the gorgeous gold and black Monarch, you could say the zoo offers a sanctuary for these majestic migrators. The Monarchs flutter to a 400-square-foot butterfly habitat by the Herpetarium and a larger 1,000-square-foot plot outside the zebra exhibit. It's the smaller one, though, where you'll find the most Monarchs during the summer. The tiny butterfly patch is planted with Mist Flowers—white and purple—Turk's Cap, Lantana, Verbena, Tropical Milkweed and Hummingbird Bush. Watch for these birds too, as they flit among the flowers with their fluttering friends. Look for a walk-through butterfly area in years to come.

Hours: 9 a.m. to 5 p.m. daily, closed Thanksgiving Day, Christmas Day and New Years Day.

Admission: $3 adults, $2 seniors (60+) and ages 3-12, free for ages under 3.

Texas Wild: Animals Alive!

Witte Museum

3801 Broadway, San Antonio, TX 78209
210-357-1900
www.wittemuseum.org

San Antonio's oldest museum is buzzing with activity and it's not just the Honey Bees causing all the commotion. This crowd pleasing permanent exhibit has an odd assortment of live critters illustrating the variety of wildlife in the Lone Star state, including Lucy & Ethel, the cutest pair of Black-Tailed Prairie Dogs you'll ever meet.

While these two friendly furry creatures stand up to greet you, you'll have to look closely to find the scorpions that use camouflage to hide from their insect prey. Two large hairy tarantulas are easy to spot, as is the Giant Centipede. The observation honey hive draws quite a crowd, so you may have to wait a few minutes to get a bug's-eye view of the honey-making activity.

Joining this collection of invertebrates are slithering snakes such as the Rat Snake that can reach a length of 80 inches, Grasshopper Mice, Green Tree Frogs, a Spiny Lizard, a color-changing Mediterranean Gecko and freshwater fish from the San Antonio River.

A hands-on learning experience, the exhibit's displays are low enough for kids to easily see and not so high that adults have to crawl on the floor. You may want to *sit* on the cushy rubber flooring however, during storytelling sessions. The exhibit area has floor puzzles, a backyard bugs game, insects that you examine with magnifying glasses and a touch-screen computer.

Hours: 10 a.m. to 5 p.m. Mon. and Wed.–Sat., 10 a.m. to 9 p.m. Tue., Noon to 5 p.m. Sun. Closed third Monday in October, Thanksgiving, Christmas Eve and Christmas Day. Call for summer hours.

Admission: $5.95 adults, $4.95 seniors (65+), $3.95 ages 4–11, free for ages 3 and under and free on Tue. 3 p.m. to 9 p.m.

Le Chateau des Papillons

Scarborough Faire the Renaissance Festival

P.O. Box 2914, Waxahachie, TX 75168
www.ButterflyChateau.com
www.ScarboroughRenFest.com

Like an emerging butterfly, **The House of Butterflies** (the English translation of the name), opens its beauty at **Scarborough Faire the Renaissance Festival** every year. The 1,260-square-foot butterfly house and boutique is the first of its kind at a renaissance festival, and to the delight of any papillon enthusiast, features native Texas species.

About 300 delicate jewels dance in the greenhouse type structure landscaped with native Texas plants such as Eve's Necklace, Texas Indigo Bush, Anacacho Orchid trees and Texas Persimmons. Among the 26 butterfly varieties: American Ladies, Hackberry Emperors, Goatweed Leafwings, Checkered Whites, Gorgone Checkerspots and Southern Dogface Butterflies.

Stroll on the interior path and you'll come across Desert Willows, Clasping Coneflowers, Texas Yellow Stars, Cutleaf Daisies, Honeysuckle and Mexican Hats planted along the way. Fluttering among these butterfly-friendly plants are Question Marks, Sleepy Oranges, Mourning Cloaks, Buckeyes, Monarchs, Queens, Red Admirals, Zebra Longwings and Viceroys. The big five Swallowtails are all here too: Giants, Pipevines, Tigers, Zebras and Eastern Blacks.

> A hidden caterpillar makes the Mexican Jumping Bean jump.

Also look for butterflies on dishes of rotting fruits such as mango, banana and cantaloupe placed throughout the house. The fruit sugar provides extra nourishment. The butterfly habitat also features an emergence enclosure in one corner, where lucky visitors can see an adult butterfly exit from its chrysalis.

Before leaving the butterflies behind, be sure to visit the boutique filled with elegant garden keepsakes and other butterfly-themed treasures.

Hours: 10 a.m. to 7 p.m. weekends including Monday Memorial Day, mid-April to first weekend in June.

Admission: $2 adults, $1 ages 12 and under. Does not include gate admission to the faire.

The Ballet of the Butterflies

Blachly Conservatory
Texas Discovery Gardens

Fair Park, 3601 Martin Luther King Blvd., Dallas, TX 75210
214-428-7476
www.texasdiscoverygardens.org

For 24 days bursts of butterflies bloom at the state's second oldest botanical institution. The event is part of the **State Fair of Texas** (www.bigtex.com), which spreads itself out in **Fair Park**, home to the gardens and several big educational centers including the **Dallas Museum of Natural History**.

About 600 to 800 native and exotic butterflies swirl among the flowering nectar plants under the 7,200-square-foot conservatory's glass roof. With permits for up to 119 different species, you're sure to see some stunning shimmering wings. Among the most common sights: electric Blue Morphos, Zebra and Tiger Longwings, Julias, Postmen, Crackers, Paper Kites, Clippers, Cattlehearts, Question Marks and Common and Great Mormons. See some of these brilliant beauties in their pupa stage in a glass emergence case that lets you observe the last stages of metamorphosis as adults shed their chrysalides and unfold their new wings.

Monarchs are a definite at the temporary butterfly habitat, which also features an educational display on these migrating marvels. The Monarchs 3,000-mile migration path from Canada to Mexico includes Texas. Learn just how high the butterflies must fly to make this incredible journey and why they are such a vibrant orange. Look for daily Monarch tagging and releases as part of **Monarch Watch**, an international effort to track their migration.

Future plans include a new permanent butterfly house and insectarium, due to a $2.5 million grant from the Rosine Foundation Fund. This will replace the **Blachly Conservatory** and live butterflies will be seen year-round.

Hours: 10 a.m. to 5:30 p.m. last Fri. in September to third Sun. of October.

Admission: (Live butterfly display) $4 adults, $3 seniors and ages 5–12, free for ages 4 and under.

Vermont

Montshire Museum of Science

One Montshire Rd., Norwich, VT 05055
802-649-2200
www.montshire.org

Bees and ants are busy working at this hands-on museum named for the last syllables in Vermont and New Hampshire. Central American Leaf Cutter Ants transport tiny pieces of leaves on their backs and tend to their fungus farm in this exhibit. Watch the activity through several clear Plexiglas containers. The ants are called Nature's Gardeners because they feed on the mold that grows on the mulched up leaves not on the leaves themselves. This said, observe the ants snatch up the leaves during daily "ant feedings" where handfuls of leaves are tossed into the exhibit. The fresh leaves are collected during the summer and frozen so the ants will have enough to last through a New England winter.

Discover how Honey Bees make that golden nectar through an observation hive located in the same place as the ants in the insect section of the second floor exhibit gallery. Listen to the buzz of the bees through a clear Plexiglas tube that the busy bugs use to go to and from the museum. There are tiny holes in the tube so the bees really do make quite a buzz!

Joining these live bugs are new displays of 50 pinned species of Northern New England dragonflies and damselflies and eventually the museum aims to have examples of all 104 species. Look for specimen cases featuring 75 species of butterflies and moths, 75 species of beetles and all nine local species of fireflies. Interactive elements let you simulate the flash patterns of five of these fireflies—each has a specific lighting pattern to attract mates. Three moveable magnifying glasses let you get a bug's eye view of many of the specimens, including wings, antennae and coloring.

Found any interesting bugs in your backyard lately? Save them and take them to the **Bring Your Own Bug** exhibit in the **Experiment Gallery**. There's a large kid-friendly microscope on permanent display that you can use to magnify your backyard find. Be sure to look up at the ceiling to see the six-foot-long model of a Monarch hanging overhead.

Located on 110-acres of woodland surrounded by a three-mile network of nature trails, don't miss spending some time soaking up this natural haven. The trails border the Connecticut River and there are picnic areas for a nice lunchtime break. Check in at the front desk before you take a trek.

In July 2002 the museum opens an intriguing outdoor **Science Park** with dozens of exhibits that tap into the surrounding environment.

Hours: 10 a.m. to 5 p.m. daily. Closed New Year's Day, Thanksgiving and Christmas.

Admission: $6.50 adults, $5.50 ages 3–17, free for ages under 3.

John Hampson's Bug Art

Fairbanks Museum & Planetarium

1302 Main St., St. Johnsbury, VT 05819
802-748-2372
www.fairbanksmuseum.org

Okay, so there are no live bugs here. But the dead ones are so fascinating that you just have to stop by if you're anywhere near this town close to the New Hampshire border. **Bug Art** is one of the museum's most popular exhibits and it's no wonder. Who would have ever thought of using bugs to design a mosaic of Abraham Lincoln? John Hampson did.

An English immigrant and former employee of Thomas Edison, Hampson used about 7,000 insects to create nine two-foot square portraits of historic figures including George Washington and General Pershing. The works of art stem from Hampson's interest in entomology while living in Newark, NJ, during the 1870s. Other artworks portray images of the American Flag and patchwork quilt designs such as the North Star and Centennial Wheel. Look closely and you'll see that the intricate designs are comprised of bug wings, legs and torsos. An interesting fact about Hampson—he designed a phonograph while working with Edison but quit when Edison made milking a cow part of the job duties.

Besides **Bug Art**, the museum founded by industrialist Frederick Fairbanks in 1891, has 18,000 square feet of exhibit space and includes a comprehensive collection of Northern New England's birds and wildlife, a **Weather Station** and Vermont's only **Planetarium**.

Hours: 9 a.m. to 5 p.m. Mon.–Sat., 1 p.m. to 5 p.m. Sun.

Admission: $5 adults; $4 seniors; $3 ages 5 to 17; free for ages under 5; $12 families with maximum 3 adults, no limit on children.

Virginia

Butterfly Station

Danville Science Center

677 Craghead St., Danville, VA 24541
(434) 791-5160
www.smv.org/wdanvil.html

Opened for its first full season in 2000, this mesh-covered butterfly house has been an inspiration to the local community. Other area gardens are following the science center's lead by planting plots that attract native butterflies. Danville City Beautiful even plans to create a butterfly garden at the **Ballou Park Nature Center**.

The **Butterfly Station** is a community project, with volunteers who help with the weeding and planting and local businesses that donate plants, tools, and supplies. People are always welcome to spend their time helping in the garden. The station also accepts donations of native butterflies, moths, turtles, dragonflies, ladybugs, tadpoles and frogs. Click on the Lepidoptera Newsletter on the center's Web site for the garden's wish list, gardening tips and to contact **Butterfly Station**'s manager.

The science center, a sister museum to the **Science Museum of Virginia**, promotes a hands-on environment for its visitors to the butterfly garden. You can take a guided tour of the station to learn what local plants attract and feed butterflies and their caterpillar selves, map a Monarch migration route or mix up some **Mung Juice**. (The recipe for this butterfly nectar appears at the end of this profile.) Kids can pretend to be a caterpillar changing into a butterfly by crawling into a sleeping bag chrysalis with towel wings. Discover wriggling wonders in the garden's worm farm underneath a bamboo teepee and see how compost enriches a garden.

So what about the wings of beauty at **Butterfly Station**? You'll see a variety of local species depending upon the day and how well each type is doing in a given year: Eastern Commas, Question Marks, Red Spotted Purples, Silver Spotted Skippers, Spring Azures, Eastern Railed Blues, Luna Moths, Hummingbird Moths, Red Admirals, Zebra Swallowtails, Gray Hairstreaks and more. You'll see about 30 butterflies and moths, the same number of caterpillars and 30 to 50 chrysalides. Placed on hatching boards and hung throughout the greenhouse, you can watch the butterflies emerge from their temporary homes. You'll also see a few moth cocoons brought in from local residents.

The 34' by 66' structure is filled with a mix of host and nectar plants: Yarrow, Coneflower, Joe-Pye Weed, Lavender, Asters, Parsley, Dill, Milkweed, Lantana, Verbena, Coreopsis, Thistle and others. Cherry, Poplar, Paw Paw, Willow and Sassafras trees are also found inside. There's a small pond with tadpoles, frogs and Koi and a waterfall area at the end of the greenhouse. Outside the enclosure the landscape includes a butterfly garden area, a pergola and benches.

Mung Juice

Start with a base of molasses; add old bananas or other overly ripe fruit. Mix with enough beer so that the concoction spreads easily. Let the mixture ripen in a container for a few days then spread it onto a tree limb or other area that can become a natural feeding station. The homemade butterfly brew will lure species like Anglewings, Satyrs and other species that are not attracted to flowers.

Hours: 9:30 a.m. to 5 p.m. Tue.–Sat. Closed Sunday, Monday, Thanksgiving Day and Christmas Day.
Admission: $4 adults, $3 seniors (60+) and ages 4–12, free for ages 3 and under.

Bioscape

Science Museum of Virginia

Historic Broad St. Station, 2500 West Broad St., Richmond, VA 23220
800-659-1727
www.smv.org

Who knows what creepy crawly creatures you may find when you visit the museum's **Bioscape Lab 2**. Inside **Gallery Education** on the second floor, the **Arthropods, Arachnids and Insects Cart** is brimming with scuttling critters. Learn what's an insect and what's not when you compare a millipede to a Praying Mantis. Visit with a Flat-Rock Scorpion (arthropod) and a Pink Toe Tarantula (arachnid). Watch Eastern Toe Biters and Giant Water Bugs swim. In the past, the cart's been home to grasshoppers, crickets, Madagascar Hissing Cockroaches, Bess Beetles, Boll Weevils and Tobacco Hornworms. If the cart isn't open when you visit, be sure to look through the lab's glass windows where the critters are always visible. For an extra dose of bugs visit the mounted butterfly and moth displays in the **Biodiversity Exhibit**.

Hours: 9:30 a.m. to 5 p.m. Mon.–Sat., 11:30 a.m. to 5 p.m. Sun.

Admission: $6 adults, $5.50 seniors (60+), $5 ages 4–12, free for ages 3 and under.

Butterfly Garden

Virginia Living Museum

524 J. Clyde Morris Blvd., Newport News, VA 23601
757-595-1900
www.valivingmuseum.org

Welcome to one wild museum! This truly is a living museum, teeming with wildlife from bobcats to beavers to Bald Eagles. And of course, bugs and butterflies. Located in a wooded area alongside **Deer Park Lake**, the museum boasts a 20,000-square-foot building for indoor exhibits including a 60-foot-long aquatic-life diorama of Virginia's **James River** from the mountains to the sea, a touch tank and discovery center. A half-mile boardwalk winds through a 13-acre forested site filled with live animals exhibited in natural habitats.

You'll find plenty of flying flowers blooming at the museum's 1,500-square-foot butterfly garden, a native wildflower meadow and children's learning garden. Monarchs, Sulphurs, Swallowtails, Painted Ladies, Fritillaries and many other butterflies are attracted to the gardens. Discover what plants entice these colorful creatures to sip nectar and what plants are needed for the larval stage. Learn how you, too, can create a backyard haven for wildlife.

In late summer and early fall follow all the stages of a butterfly's life from caterpillar to pupa to emerging adult butterfly. The museum erects special rearing chambers for Monarchs, which are then tagged and released to migrate to Mexico for the winter.

Butterflies aren't the only incredible insects you'll see. Check out the live Betsy Beetles, the working beehive and displays of preserved beetles and butterflies inside the museum.

Until November 2002, the museum's changing exhibit gallery features a **Pollination Station** where visitors can explore the importance of pollination through interactive displays. Discover how bees and butterflies pollinate flowers and watch butterflies emerge from their chrysalides in the **Butterfly Nursery**. Kids can *bee* a pollinator when they interact with the **Pollinator Puppet Stage**.

In 2004 a new 62,000-square-foot museum building will open with Cypress Swamp and Appalachian Mountain habitats.

Hours: (Summer) 9 a.m. to 6 p.m. Memorial Day to Labor Day. (Winter) 9 a.m. to 5 p.m. Mon.–Sat., Noon to 5 p.m. Sun., Labor Day to Memorial Day.

Admission: $7 adults, $5 ages 3–12, free for ages 2 and under.

Bristow Butterfly Garden

Norfolk Botanical Garden

6700 Azalea Garden Rd., Norfolk, Virginia 25318
757-441-5830
www.virginiagarden.org

Inside this 155-acre botanical oasis is a 1.5-acre habitat brimming with butterfly-friendly plants. Surrounded by **Lake Whitehurst**, the butterfly garden starts to come alive in April and May depending upon the weather. Master Gardener Linda Little has identified 32 different varieties of butterflies that have visited the gardens' abundant flora. Among them: Black Swallowtails, Palamedes Swallowtails, Spicebushes, Tiger Swallowtails, Red Spotted Purples, American Ladies, Painted Ladies, Question Marks, Pearl Crescents, Variegated Fritillaries, Common Wood Nymphs, Long Tailed Skippers, Silver Spotted Skippers and Monarchs. The garden sprouts a large collection of Buddleia "butterfly bush" which butterflies love. Perennial plants include Sedums, Veronica, Coreopsis and Verbena while annuals like Petunias; Zinnias, Gompherena and Vinca are planted in spring.

With plenty of host plants like Fennel, Parsley, Hops, Senna, Passion Vine and Sassafras, you'll get to see all the phases of butterfly life particularly in July and August, the peak time for the garden's butterfly activity. An adjacent meadow is home to host plants not typically found in a garden setting such as Plantain, Dock and Clover, so the area is bursting with butterflies. Sit for a spell on the benches and view the butterflies or watch as the kids run through a 200' by 150' butterfly shaped maze that's mowed into the meadow.

A fun fact about the **Norfolk Botanical Garden**—it's the nation's only botanical garden that can be toured by both tram and boat.

Hours: 9 a.m. to 7 p.m. mid-April to mid-October; 9 a.m. to 5 p.m. mid-October to mid-April.

Admission: $6 adults, $5 seniors, $4 ages 6–16, free for ages 5 and younger with a paying adult.

Washington

Tropical Butterfly House and Insect Village

Pacific Science Center

200 Second Ave. North, Seattle, WA 98109
206-443-2001
www.pacsci.org

It may be cold and rainy outside but inside the first U.S. museum founded as a science and technology center, Costa Rican Blue Morphos, Malaysian Peacock Pansies and Filipino Paper Kites are flying in a balmy 80 degrees, 80 percent humidity. The butterflies bloom in the 4,000-square-foot glass-enclosed butterfly habitat in the **Ackerley Family Exhibit Gallery**. Due to Seattle's latitude, a skylight nearly half-the size of the floor area and 28,000 watts (5,000 to 7,000 candles) of full-spectrum artificial light help create the lush living area for these tropical beauties.

Rain forest plants and trees create the perfect place to nurture adult butterflies: Flowering Maples, Butterfly Bush, Chinese Hibiscus, Trumpet Vines and Star Clusters. Stroll down a garden path to see some of the 50 exotic species the science center is permitted to release in the butterfly house. These free-flying butterflies are displayed on a rotating basis. Among them, flame-bordered Charaxes, Forest Queens and Checkered Limes from Kenya; Zebra Mosaics, Malachites and Isabella Tigers from Costa Rica; Roses and Scarlet Mormons from the Philippines; Autumn Leafs, Tailed Jays, and Atlas Moths from Malaysia and Cattle Hearts from Ecuador. Watch through a window to see butterflies and moths emerge from their chrysalides and cocoons, imported from their countries of origin. Look for informational displays on metamorphosis, tropical rainforests and how to distinguish a moth from a butterfly.

Bug buffs can explore the other bugfully delightful habitat at the museum, its innovative **Insect Village**. Discover the structures that eusocial bugs call home from a termite mound to an anthill to a very busy beehive. Pet your fave arthropod including Madagascar Hissing Cockroaches, walking sticks, Giant Centipedes and a hairy tarantula at the **Insect Zoo**. Staff and volunteers help you with the live critters, answer questions and give daily demonstrations. Enter the big bug top when you step inside the **Insecta Sideshow**. Encounter fleas, the champion jumpers; Herculean ants; Masters of Disguise like the Io Moth; the Amazing Aqua-Beetle and bugs that emit secret codes. In the **Insects 101** exhibit, visit an entomologist's field tent to learn the basis of bug biology or check out the **Wall of Fame** for trophies displaying information on the biggest, the smallest and heaviest bugs. There's more when you visit the village's freestanding interactive stations: try on different antennae, listen to a cricket concerto and learn why Madagascar Cockroaches hiss.

Hours: (Winter) 10 a.m. to 5 p.m. Mon.- Fri., 10 a.m. to 6 p.m. Sat.–Sun. (Summer) 10 a.m. to 6 p.m. mid-June through Labor Day.

Admission: $8 adults, $5.50 seniors (65+) and ages 3–13, free for under age 3.

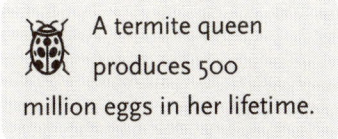

> A termite queen produces 500 million eggs in her lifetime.

Bug World and Butterflies & Blooms

Woodland Park Zoo

5500 Phinney Ave., North Seattle, WA 98103
206-684-4800
www.zoo.org

Go around the world in 80 minutes or less when fictional entomologist Mariposa takes you on a tour of **Bug World** located in the zoo's **Temperate Forest** zone. The 960-square-foot indoor exhibit features 12 environmentally diverse arthropod exhibits: Sow Bugs, American Cockroaches in a simulated kitchen and Assassin Bugs in an African savannah setting. Look for these two tanks: the **Desert by Day** with katydids and Darkling Beetles and the **Desert by Night** with Dung Beetles and a Hairy Scorpion with a black light option so you can see it glow in the dark. Aquatic invertebrates include Water Bugs and Diving Beetles and a tank of crayfish. Examine Madagascar Hissing Cockroaches, Pinkwing Walking Sticks, a Bird Eating Tarantula, Peruvian Fire Sticks, a Giant African Millipede, Flamboyant Flower Beetles and Pacific Dampwood Termites. Watch the Honey Bees buzz in and out of their display hive. Each decorated tank includes notes from the entomologist's logbook from her fictional travels to temperate forests, deserts, savannas and tropics.

For more bugginess, find the Leaf Cutter Ants in the award-winning **Tropical Rain Forest** exhibit. The busy bugs are right inside the entrance and the exhibit extends all the way to the simulated rain forest's canopy. The primary viewing area is at the forest floor level. Watch the ants climb up the simulated tree trunk and return with cut leaf pieces. Behind the scenes at the exhibit's top animal handlers feed the ants leaf browse. At the bottom you can observe the ants turn the leaves into a fungal farm.

From May to September, the zoo gets even buggier at its annual **Butterflies and Blooms** exhibit that spills into 17,000 square feet of space in the **North Meadow** zone. Divided into three sections, the exhibit first brings visitors to the **Life Cycle Landscape**. Here you can pretend that you're a butterfly in one of its life stages: egg, larva (caterpillar), pupa (chrysalis) or adult butterfly. You do this by putting your face into larger-than-life-sized cutouts. It's a great photo opportunity.

At the **Butterflies in Flight** section you enter a kaleidoscope of blooms bursting in 3,900 square feet of an indoor landscape framed by two greenhouse structures. Fluttering around the vibrant plant blossoms are about 1,000 flying flowers representing at least 15 North American butterfly and moth species. Among those you may are encounter are Luna Moths, Dogface Butterflies, Malachites, Question Marks, Ruddy Daggerwings, Silver Spotted Skippers, Sleepy Oranges and Spicebush Swallowtails. Classical music drifts through the butterfly habitats: woodland clearing, meadow's edge and open meadow. The diversity of plants, trees and shrubs gives you a look at various butterfly behaviors. Midway through this section, you'll see the boxes that house the butterfly pupae. Observe as butterflies emerge from their chrysalides and moths exit their cocoons. These adults will alight to a nearby tree and dry their damp wings before flying around the exhibit.

Dispersed between this area and the exhibit's 6,000-square-foot outdoor **Conservation Garden** are 7,200 trees and other plants from more than 75 varieties such as Sitka Willow Trees, Butterfly Bush, Hollyhock, Black-Eyed Susans, Torchflowers, and Passionflowers. The garden highlights the hardy plants of the Pacific Northwest and shows visitors how to attract wild butterflies to their own backyards. Questions? Several zoo horticulturalists, docents and other volunteers are ready with answers.

Near the garden you'll find an outdoor **ZooStore** stocked with butterfly gardening books, field guides and butterfly feeders.

Hours: (Zoo) 9:30 a.m. to 6 p.m. daily May 1 to September 14, 9:30 a.m. to 5 p.m. daily September 15 to October 14. (Butterflies & Blooms) Opens at 10 a.m. May to September.

Admission: (Zoo) $9.50 adults, $8.75 seniors (65+) and students with ID, $7 ages 6–17 and the disabled, $4.25 ages 3–5, free for ages 2 and under. (Butterflies & Blooms) $1 per person in addition to regular zoo admission, free for ages 2 and under.

> Woodland Park Zoo sells its poop as compost during its Zoo Doo event.

West Virginia

Joan Stifel Corson Butterfly and Wildflower Gardens

Schrader Environmental Education Center at the Oglebay Institute

Oglebay Park, Wheeling, WV 26003
304-242-6855
www.oionline.com

See nature's rainbows fly through this 15,000-square-foot butterfly and wildlife garden next to West Virginia's premier environmental center. Look for Aphrodite Fritillaries, Little Sulphurs, Meadow Fritillaries, Red Admirals, Tiger and Giant Swallowtails, Painted Ladies and the queen of them all, Monarchs. The gardens include **Butterfly Habitat Beds**, a **Backyard Butterfly Garden Demonstration Area** and a **Wildflower Meadow Garden**. Self-guided tour brochures are available for butterfly viewing and for the 4.5 mile **A.B. Brooks Discovery Trail** system at the adjacent **Schrader Environmental Center**.

See a waterfall on the **Vista Trail** and discover Bluebirds and get close-up looks at wildlife on the **Habitat Discovery Loop**. The **Hardwood Ridge Trail** features three scenic overlooks, rest areas, and maple sugaring spots, wildflowers and wildlife.

The **Schrader Environmental Center** features plenty of hands-on excitement for kids. A neat aspect is the bird watching at **Bird Café** featuring video monitoring of outdoor bird feeders. Four window feeders use one-way glass so you can press your nose against the glass and the birds won't be disturbed. Look for the historic stained glass window at the café entrance. The center is also noted for its "green architecture." Some of the environmentally sound elements of the building include recyclable steel product beams; rubber flooring made from 95 percent recycled bus and truck tires and tree products from new growth timber. Guided tours available on Saturdays in February, on Saturday and Sunday in mid-March and daily Memorial Day to Labor Day.

Hours: 10 a.m. to 5 p.m. Mon.–Sat. Open Noon–5 p.m. Sun. starting mid-March.

Admission: (Guided tours) $2 per person, $5 per family. (Self-guided tours) Free.

Good Zoo & Benedum Planetarium

Oglebay Family Resort
Route 88 North, Wheeling, WV 26003
800-624-6988
www.oglebay-resort.com/homepage.htm

Comprised of only 30 acres, this zoo is nestled in a much larger complex, the 1,650-acre **Oglebay Family Resort**. Among the hills and valleys, flowers and trees are 85 species of animals from North American bears and buffalo to Red Pandas and Tamarin Monkeys. In the indoor **Discovery Lab** you'll find a small selection of invertebrates: Giant Hissing Cockroaches, Northern Walking Sticks, Giant African Millipedes and a Red Knee Tarantula. The millipedes may be off exhibit when you come because of visits to local schools. The lab features several hands-on activities and some other fun live creatures including Poison Dart Frogs and Tiger Salamanders.

Kids will enjoy a visit to the barn filled with domestic animals such as llamas, donkeys, pigs and ponies. You can hand feed the goats and deer.

Hours: 11 a.m. to 4 p.m. daily January to March. Hours and rates vary for special events, summer and holidays.

Admission: $3.25 adults, $2.75 ages 2–17.

Wisconsin

Honey of a Museum

Honey Acres

N1557 Highway 67, Ashippun, WI 53003
800-558-7745
www.honeyacres.com

There's tons of sticky stuff to be found at this beekeeping and honey packing company located amidst sweet smelling clover and Linden trees, perfect for bees to make their lip-smacking amber liquid. Owned by the Diehnelt family since 1852, **Honey Acres** offers visitors "A romance with honey." Their museum gives you a close up look at bees, beekeeping and honey packaging. The bee tree offers a window into a honey of a hive and you can see the bees busily building their combs and tending to their young. Educational displays depict the life cycle of bees and historic exhibits describe beekeeping around the world from yesterday and today. Watch a 20-minute multi-media show, *A Honey of a Story*, then take a walk outside and get a great scenic view. Don't go without tasting some honey. The gift shop offers an assortment of honey items from mints to mustards to beeswax candles and bottles and bottles of syrupy sweets. Look closely at the bottles, which are reproductions of the 1870 Muth jar. Look, too, at the artwork on the gift shop's window.

Hours: 9 a.m. to 3:30 p.m. Mon.–Fri. year round; 12 p.m. to 4 p.m. Sat.–Sun. May 15 to October 30.

Admission: Free.

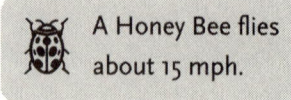

> A Honey Bee flies about 15 mph.

Puelicher Butterfly Wing

Milwaukee Public Museum

800 W. Wells St., Milwaukee, WI 53233
414-278-2702
www.mpm.edu

Be prepared for a swirl of electrifying colors when you enter this 1,800-square-foot glass-enclosed garden housing hundreds of tropical and domestic butterflies. The two-story double glass structure fronts the museum and entices everyone to come inside. The rain forest environment includes blooming plants, tranquil music and a cascading waterfall. A changing array of butterflies is exhibited with about 20 species seen at any one time. The butterfly beauties hail from Central and South America, Texas, the Philippines, Malaysia and Africa. Try to spot Idea, a Southeast Asian butterfly with wings that look like leopard skin.

Other species include Blue Morphos, Black Swallowtails, Owl Eyes, Monarchs and lots of Longwings.

The *Milwaukee Journal Sentinel* newspaper has a great online article with full color photographs of the museum's butterfly species at http://www.jsonline.com/enter/planit/jun00/butterfly.asp. Print it out and take it with you. You could even turn your visit into a butterfly scavenger hunt.

A **Transformation Station** inside the butterfly wing lets you watch as the butterflies emerge from their chrysalides. Look for "scrubbies," yellow and red spongy objects hanging through out the enclosure. Sprayed with sugar water, these supplement the nectar for the bevy of butterflies.

The butterfly experience doesn't end here. The 3,000-square-foot gallery adjacent to the **Butterfly Wing** includes a working **Butterfly Laboratory**. Visitors can study butterflies and moths and raise their "Insect IQ" with hands-on activities and a video. Very young children enjoy roaming through a super-sized butterfly garden where they can wear costumes and transform into caterpillars or butterflies. A Wisconsin backyard diorama helps local visitors learn how to attract these winged wonders to their own backyards. The **Wall of Diversity** wraps around the gallery and showcases more than 1,000 different species from the museum's Lepidoptera collection, which dates back to the 19th Century.

For more butterfly beauty head for the museum's **Rain Forest** exhibit on the first floor. Try to find the hundreds of tropical butterflies on display while learning about biodiversity and how scientists are trying to save this life on Earth.

Hours: 9 a.m. to 5 p.m. daily. Closed July 4, Thanksgiving and Christmas.

Admission: $6.50 adults, $5 seniors (60+), $4 ages 4–17, free for ages 3 and younger.

Blooming Butterflies

Olbrich Botanical Gardens

3330 Atwood Ave., Madison, WI 53704
608-246-4550
www.olbrich.org

It only lasts from mid-July to late August but during this six-week exhibit these botanical gardens definitely bloom with butterflies. Up to 2,000 butterflies and moths can be found in the 10,000-square-foot pyramid shaped **Bolz Conservatory**. About 25 species of butterflies and several large moth species hail from Wisconsin, Texas and sub-tropical Florida. Inside the two-story glass structure the butterflies emerge from their hatchery: Malachites, White Peacocks, Zebra Longwings, Viceroys, Red Admirals as well as huge Cecropia moths. About 1,000 fleeting flyers are released into the conservatory each week. The conservatory is big enough to make space for the butterflies and free-flying birds. The birds are not predators since they are seed-eating varieties: canaries, Waxbills and Diamond Doves. There are hundreds of exotic plants and flowers, too: 500 sweet-smelling blooming Orchids, 30 different Palm trees and various tree perching plants known as Epiphytes.

In a corner of the pyramid you'll find the rustic **Exploration Station**, a thatched research hut where kids can learn about the rain forest ecosystem. On weekends the **Snipping Gardens** are open—kids can make plant cuttings and bring them home to start their own mini-conservatories.

There's also a **Butterfly Walk** in the outdoor gardens. Take a self-guided tour and discover what flowers and plants attract native Wisconsin butterflies, which you should see flying in the open air.

Hours: (Conservatory) 10 a.m. to 4 p.m., Mon.–Sat; 10 a.m. to 5 p.m. Sun.(Outdoor Gardens) 9 a.m. to 4 p.m. daily, 8 a.m. to 8 p.m. April to September. Closed Thanksgiving and Christmas.

Admission: (Conservatory) $1 per person, free for ages 5 and under, free on Wednesday and Saturday mornings 10 a.m. to 12 p.m. (Outdoor Gardens) Free.

Mosquito Hill Nature Center Butterfly House

N3880 Rogers Rd., New London, WI 54961
920-779-6433

Watch the colors of Wisconsin wing their way through this temporary butterfly enclosure from early July to late August. A 1,500-square-foot screened structure spreads over a garden planted with hundreds of butterfly-attracting perennials, annuals, shrubs and trees: Pearly Everlastings, Pussy Toes, Candy Tufts and Blazing Stars and Black-Eyed Susans. The native butterflies are all raised in captivity or temporarily removed from the wild: Pearl Crescents, Red Admirals, Painted Ladies, Viceroys, Mourning Cloaks and Little Wood Satyrs. Thirty well-trained volunteers staff the butterfly house and can answer your questions about the fluttering beauties or about the plants that attract them. Ask them how to start your own butterfly garden; the nature center has supplies and butterfly-themed gifts.

Hours: 11 a.m. to 3 p.m. Wed., Sat., and Sun.

Admission: $1 donation per visitor requested.

Discovery Center

Henry Vilas Zoo
702 South Randall Ave., Madison, WI 53715
www.vilaszoo.org

Don't let the giant Black Widow hanging from the ceiling of this hands-on center scare you away. It's only a sculpture made from painted greenhouse foam and corn stalks. Inside, the 28-acre zoo's **Discovery Center** has a nice invertebrate section with live crawlers housed in glass terrariums: Giant African Millipedes, South Vietnamese Walking Sticks, Madagascar Hissing Cockroaches, an African Scorpion and a tarantula. Aquarium tanks feature aquatic invertebrates: Brittle Stars, Coral, Star Fish, snails and urchins.

The invertebrate homes sit on kid-friendly tables with magnifiers. Glass specimen cases that were confiscated because they contained illegal mounted butterflies and large beetles like the Rhinoceros and Hercules species, are also on view. Four large insect head models let visitors examine the mouthparts of a mosquito, butterfly, House Fly and beetle.

Four other tables provide an up close look at other live critters. The mammal table has a real hedgehog, the bird table features a little quail, the reptile table has snakes and Leopard Geckos and the amphibian table houses Fire Belly Toads and Poison Dart Frogs. A fifth table holds a huge whale skull and a variety of teeth for visitors to examine including those from whales, sharks and humans.

The **Discovery Center** once had an observation bee hive but the colony died out. Look for one in the future as well as the **Tropical Forest Aviary**, which is expected to include live insects. And if you visit in the summer, be sure to catch a free camel ride most Sundays in the late morning.

Hours: (Main Zoo) 9:30 a.m. to 8 p.m. June through Labor Day weekend, 9:30 a.m. to 5 p.m. September through May. Closed afternoons on Thanksgiving Day and the day after; Christmas Eve and Christmas Day; New Year's Eve and New Year's Day; and Martin Luther King Day. (Children's Zoo) Open Memorial Day weekend through Labor Day.
Admission: Free.

Wyoming

Insect Gallery

University of Wyoming Insect Museum

Room 4018, Agriculture Building, Laramie, WY 82071

307-766-2298

www.uwyo.edu/ag/psisci/braconid/museum.htm

This may be the only place in Wyoming where you can see live bugs. But what better place than this university research museum, which houses the largest insect specimen collection in the state—250,000 of them. The **Insect Gallery** was created in 1993 to celebrate 100 years of Wyoming entomology and to serve as place where the public could learn about bugs.

The small insect zoo has the typical live arthropods on display: Madagascar Hissing Cockroaches, Giant Millipedes, tarantulas, crickets and Bess Beetles. What's fun is the 50-gallon tank depicting a western village—its residents are the hissing cockroaches. It's popular with the kids so be sure to reserve enough time for them to watch for a while.

The rest of the gallery includes well-done educational displays of exotic and local insect specimens, mostly butterflies, moths, beetles and stick insects from Wyoming, Colorado, Africa, South and Central America, and Australia. The gallery is kid friendly with about half of all the displays at kid level. There are insect artifacts, a discovery cabinet, bug models and a kids' book corner with insect books. Kids love to use the hand stamps to tattoo themselves with bug pictures and can search for insects in the tropical mural painted by student artists.

Curator of the gallery, Dr. Scott R. Shaw, is your contact to arrange for school and large group tours at 307-766-2298. Otherwise, visitors should call the gallery to check on public viewing hours: 307-766-2298. You can make the visit a

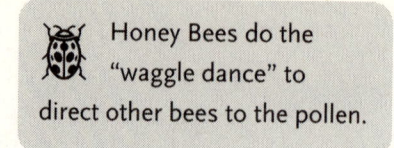

> Honey Bees do the "waggle dance" to direct other bees to the pollen.

complete day trip for the family by visiting one or more of the 780-acre university's other museums: the **UW Geological Museum**, **UW Rocky Mountain Herbarium**, **UW Planetarium** and the **UW Anthropology Museum**.

Hours: Generally 9 a.m. to 4 p.m. weekdays.

Admission: Free but tax-deductible donations are appreciated.

CHAPTER TWO

Bugged Out

**A swarm of bug festivals
and other infested events**

● ● ● ● ● ●

Go ahead, bug me! There's no better fun than getting bug-eyed at an event swarming with live bugs, infested with insect enthusiasts, and crawling with creative critter activities. Most of the events listed here are annual and are more popular each year.

Many of these insect fests are one-day events while others are held only for an afternoon. It's a blast to go buggy at a bug day and you may wish you could get bugged more. So plan a visit to a different festival in another city or state.

February

Insect Fear Film Festival

University of Illinois at Urbana-Champaign
Urbana, IL

"**B**ee afraid, Bee VERY Afraid!" That was the theme for the **University of Illinois**' creepiest annual event, the 2000 **Insect Fear Film Festival**. Originally spawned in 1984 by Entomology Department professor May Berenbaum, this festival has been reproducing every year since. Each film fest includes two or three feature-length movies like *The Fly*, *Mothra*, and *Them*. The 12-hour festival includes film or TV shorts like "ZZZZZZ," a 1964 *Outer Limits* episode. The festival is not just about staying glued to your seat. The evening event includes a visit to the arthropod petting zoo in the lobby, snacking on insect treats such as deep-fried waxworms and meeting bug experts like a bee wrangler or tarantula handler.

Contact: Department of Entomology at 217-333-2910
www.life.uiuc.edu/Entomology/home.html

Hug-a-Bug

Houston Museum of Natural History
Houston, TX

Spread the love of a ladybug every Valentine's weekend by joining in the release of thousands of these beneficial insects. Each child who visits the butterfly center this weekend gets a vial of live ladies to release inside the center. The ladybugs are a butterfly-friendly way of controlling pest insects inside the tropical rain forest environment.

Kids also receive a colorful ladybug balloon while women get a rose.

Contact: 713-639-4629
www.hmns.org

Iceworm Festival
Cordova, Alaska

Tucked away in the Alaskan wilderness but within an hour's plane ride from Anchorage, Cordova is a working town on Orca Inlet at the end of Prince William Sound. It's a bit remote but not so far that you can't join the town's residents in celebrating the region's mysterious iceworm. This tiny elusive creature squirms between ice crystals in temperate glaciers.

The town turns out quite a party for the worm with everything from an iceworm breakfast to an iceworm dart contest to an historical iceworm display. Watch the **Blessing of the Fleet**, the **Survival Suit Race** or strap on your skis for the race in Heney Meadows. Enjoy the variety of food from an ice cream feed to oyster shucking to a clam chowder cook-off. The main event though, is the **Iceworm Parade** in downtown Cordova. Look for Miss Iceworm, a local young lady who's really earned her crown.

Contact: 907-424-7260
iceworm@ptialaska.net

March

Butterflies & Orchids
San Diego Wild Animal Park
Escondido, CA

Both the butterflies and the orchids are in full bloom during a two-week kaleidoscope of color at the 1,800-acre wild animal park's celebration of spring. Thousands of tropical butterflies are released into the lush 35-foot-high **Hidden Jungle Exhibit**, which serves as a walk-through aviary and greenhouse. See Wood Nymphs from Malaysia, yellow and black Zebras, metallic Blue Morphos, Red Cattlehearts and Owls from South America fly freely through the permanent enclosure. Discover how butterflies emerge from their chrysalides by observing them in a pupa display. Find hundreds of purple, pink, white and yellow orchids flowering under a covered walkway near **Nairobi Village**. The event starts the last week of March and goes through the first week of April.

Contact: 760-796-5621
www.wildanimalpark.com

Family Festival Bug Day

Dallas Museum of Natural History
Dallas, TX

Get close to Carl's Creepy Crawlies, watch Fire Ants in their underground colonies without getting stung and get a magnified look at lagoon insects. See the world through a bug's eyes, snack on chocolate-covered bugs and get a butterfly tattooed on your arm. The festival features several creepy crawly crafts, games and activities.

Contact: 214-421-3466
www.dallasdino.org

A queen Leaf Cutter Ant mates only once but produces 300 million babies.

April

Bug Bowl

Purdue University
West Lafayette, IN

How far can you spit a dead cricket? Find out at the largest bug bash in the Midwest where cricket spitting is an annual sport. If you expect to win at this event, which was invented here, be prepared to reach distances of at least 30 feet in the senior men's competition in order to make the *Guinness Book of World Records*. Dan Capps of Madison, Wisconsin holds the world's record of 32 feet and 1.5 inches, although he's unofficially reached distances well beyond that. If you're not up to popping chirpers across a finish line, then try sucking on a cricket lollipop or munching on caterpillar-laced chocolate chip cookies.

Crunch your buggy delights while cheering on your favorite antennaed friend at **Roach Hill Downs**. Official jockeys are picked from the youngsters in the audience. Kids can also participate in the six-legged **Caterpillar Canter**. Enter the popular cake decorating contest—cakes must feature an insect motif or shape but do *not* have to include bugs in the recipe. Pet your pals at the live **Insect Zoo** featuring Giant Madagascar Hissing Cockroaches and Guinea Stick Insects. Observe bees in a hive then sample their honey, watch butterflies fly and look at bug parts under microscopes. About 12,000 bug fans attend each year.

Contact: Tom Turpin at 765-494-4554
www.entm.purdue.edu
bug_bowl@entm.purdue.edu

Butterfly Adoption & Eco-Arts Festival

KidSpace Museum
Pasadena, CA

This small hands-on children's museum teaches kids about metamorphosis in a big way—it hands out 1,000 caterpillars to children who agree to raise them into butterflies. Children and their families sign a pledge that by adopting the caterpillars they will care for them and bring them back two weeks later as Painted Lady Butterflies. Kids make shoebox butterfly bungalows to raise the caterpillars in. They can observe the larva as they munch on the food the museum provides. Then during the museum's **Eco-Arts Festival** the children return to release their butterflies during a special ceremony.

Contact: 626-449-9144

BugFest 2002

Montana State University
Bozeman, MT

So who's responsible for cleaning up manure in South Africa, Swaziland and Namibia? The African Dung Beetle of course! You can meet these cleaners of the Earth at **Montana U.**'s insect festival held every other year. It's a weeklong fest buzzing with live bugs and butterflies plus hundreds of beetles, wasps, flies and other specimens from the university's **College of Agriculture** and its **Entomology Department** which celebrated 100 years of bug collecting in 1998. The live ones include walking sticks, millipedes and other arthropods from Belize, Malaysia, Costa Rica, Florida and Texas.

Contact: 406-994-2470
www.montana.edu

Insect Festival

University of Arkansas
Fayetteville, AR

The bees are buzzing and the cockroach races are on at the annual infested festival by the university's **Department of Entomology**. Come see what the entomologists have been raising for the past year so kids can visit a live arthropod zoo. Among the lively crew of creepy crawlies: exotic tarantulas, scorpions, centipedes and Madagascar Hissing Cockroaches that you can pet or cheer on in a race.

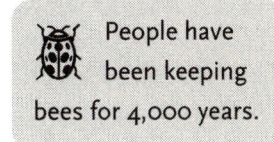

> People have been keeping bees for 4,000 years.

Buzz on over to both the Bumble Bee colonies and the Honey Bee observation hive and discover what beekeepers wear to work and how honey is extracted. Have a bugfully delightful time creating butterfly crafts, visiting with forest bugs and checking out a cotton exhibit complete with a cotton patch and cotton gin. Microscope displays provide a magnified look at all types of exotic bugs and the university's **Arthropod Museum** brings out dozens of display cases filled with hundreds of mounted flies, dragonflies, butterflies and beetles.

Contact: 479-575-4795
jbarnes@uark.edu

Bugz All Day

Lexington Children's Museum
Lexington, KY

Marvel at the bristling legs of a Giant African Millipede or stroke a seven-inch-long walking stick in this annual close encounter with creepy crawlies. Meet a Madagascar Hissing Cockroach from the **University of Kentucky**'s touchable **Bug Lab**, **Lexington Community College**'s trained termites and pest exterminator Tom Myers and his bug collection. "I Held a Bug" stickers are awarded to any one brave enough to handle a live bug. Learn why insects are so important to ecology and how bugs use camouflage to protect themselves. Kids can create antennae hats, House Fly and dragonfly masks or wear a temporary buggy tattoo. The fearless can take a lick of a cricket lollipop or crunch mealworm munchies.

Contact: 859-258-3256
www.lexingtonchildrensmuseum.com

Museum Days

Oregon State University
Corvallis, OR

Go buggy at this university's biannual party held in April and October to showcase its life sciences departments. Although **Museum Days** features the **Hatfield Marine Science Center**, the **Botany Department** and the **Fisheries and Wildlife Department**, the bugs rule on these days. The **Entomology Department** brings out its bug zoo and kids get to pet Madagascar Hissing Cockroaches, walking sticks and other incredible invertebrates. The idea is to really show kids that crawlies aren't so creepy. There are live snakes, amphibians and sea invertebrates that you can see and touch too. If you're lucky you may be asked to taste some grub, only it might actually be grubs or dry-roasted grasshoppers that you're munching on. Or perhaps sautéed silk worm pupae or stir-fried mealworms. The bulk of the visitors coming through for the events are local students, so you may want to attend after 2 p.m. when the crowd lessens.

Contact: 541-737-5520
roycel@bcc.orst.edu

International Flower & Garden Festival

Epcot Center, Walt Disney World
Orlando, FL

There's a whole lot of buggy stuff happening on the 300-acres of lush mouseland featuring more than 1,200 plant species and thousands of brilliant blooms during this annual six-weekend floral event. You need an official festival guide to traverse this lavishly landscaped land but here are a few hints about the bugged-out activities:

> The average ladybug eats 50 aphids a day.

At the **Outpost in the World Showcase**, join Dr. L. Bug in an effort to save the gardens from big bad bugs. Help release thousands of lucky ladybugs in the **Kids' Garden** on Fridays, Saturdays and Sundays. Beauty takes wing in **East Future World** when hundreds of butterflies are set free. The release takes place in the **Backyard Habitat Garden** on Fridays, Saturdays and Sundays.

During the "I Dig Bugs" themed weekend kids will find even more buggy activities to participate in. In 2002, Nathan Erwin, director of the **Smithsonian's Otto Orkin Insect Zoo**, takes you on a garden journey to uncover the secrets of the soil—insects that help plants grow and bugs that battle over blooms.

Contact: 407-WDW-MAGIC
www.disneyworld.com

Arcadia Insect Fair

Arboretum of Los Angeles County
Arcadia, CA

Entomological Research Publishers and the arboretum sponsor this fair that includes a special youth day for teachers and students. Enter the **Draw-A-Bug** contests, shop for deals on live and preserved bugs from butterflies to beetles to millipedes. Check out the 40 educational tables and merchandise vendors. Learn how to keep a hairy tarantula as a pet, plant a butterfly-attracting garden and just generally go buggy!

Contact: 562-862-9333
www.angelfire.com/biz/entomologybooks/

May

Insect Fair

Natural History Museum of Los Angeles County
Los Angeles, CA

Thousands of bugs and their creepy crawly cousins are on display at the largest insect event in the West. More than 10,000 visitors turn out for this two-day fest where the bugs still outnumber the people. About 50 exhibitors sell an incredible array of small creatures, live and preserved, including butterflies, beetles, spiders, scorpions, millipedes and centipedes. Looking for the pet equivalent of a fish? **Wee Beasties** specializes in exotic tarantulas, Malayan Millipedes and Giant Madagascar Pill Bugs. For the less adventurous, Dr. Sue of **Metamorphosis Enterprises** offers Chinese take-out tubs of mealworms to raise into Darkling Beetles. **BioQuip** is here with tons of buggy supplies and books and **Kathy's Critters** offers tables of buggy kits and toys.

Test your bugability at the museum's **Traveling Insect Zoo** where you can touch a live Giant African Millipede, a Madagascar Hissing Cockroach or a Giant Walking Stick. Try **Licket Cricket**'s crushed crickets dipped in chocolate or lollipops with a lickable cricket in the middle. The museum's **Curators Café** serves up Cricket Croquettes (tastes a little like crab cakes) and Chocolate Covered Ant Cookies. Haven't been bugged enough? Then check out the mounds of insect collecting equipment, and buggy toys, books and puzzles. Young bugatists™ can earn their butterfly wings in entomology with activities and crafts. Be sure to visit the museum's **Ralph M. Parsons Insect Zoo** and its summer outdoor butterfly pavilion.

Contact: 213-763-DINO
www.nhm.org

Bug Day

Randall Museum
San Francisco, CA

I don't which is better, the maggot races or mealworm munchies at this cultural and environmental museum's annual **Bug Day**. If you're lucky, the museum will have its strobe light **Insect Disco**. Stop in for a **Firefly Chat**, discover the joys of butterfly gardening and watch amazing bug-eating plants zap a fly or two. Let a walking stick crawl up your arm, zoom in on bugs with a video microscope and talk to bug experts. Beekeepers and live bees join the buzzing crowd that includes artists and storytellers. Co-sponsored by the **UC Berkeley Entomology Club** and **Butterfly Discovery Park**.
Contact: 415-554-9600
www.randallmuseum.org

Bug Day

Henry Cowell State Park
Felton, CA

Celebrate buggy life in the home of the Big Trees. Participate in the bug races, build-a-bug, go on a bug run. Watch amazing stories emerge in the **Insect Theatre**. Go to school and earn your honorary "Master's of Bugology" degree. The park is a wonderful place to spend the day or stay overnight camping (make reservations ahead) in the old growth forest with 1,800-year-old trees. The park has a nature center, gift shop and picnic area. The **San Lorenzo River** runs through the 4,000-acre park and the **Roaring Camp Big Trees Railroad** is just steps away from the parking lot.
Contact: Mountain Parks Foundation at 831-335-3174
http://dforthof.best.vwh.net/HenryCowell/mountainparks.html

BugFest

National Mall
Washington, D.C.

The **National Mall** is swarming with live crawling critters including tarantulas, cockroaches and a Bumble Bee colony. It's also buzzing with **National Museum of Natural History** scientists dressed in unique buggy costumes. Root for your favorite Madagascar Hissing Cockroach to creep across the finish line first during the **Roach Races**. Sample gourmet insect cuisine like Scorpion Scaloppini and Grasshopper Kabobs. Using hand lenses, kids can go bug hunting for hidden insects and spiders inside tree logs. Fold origami butterflies, sketch a bug or bring a real one in for entomologists to identify. Be sure to visit the **Otto Orkin Insect Zoo** on the museum's second floor.

> **Contact:** National Museum of Natural History at 202-357-2700
> www.mnh.si.edu

Bug Day

Columbus Zoo
Columbus, OH

Learn what it's like to be an entomologist in **Entomology Camp** when the zoo devotes a day to the bugs. **Ohio State University** graduate students show kids how to build a real bug collection or you can build a life-size bug out of bamboo. The Bee Lady brings her bees for show and tell and a bat expert explains how these flying mammals live on insects. Meet Spiderman and Stinger, the mascot of the Blue Jackets hockey team. Kids have fun making dragonfly tiaras and Leaf Cutter Ant hats plus watching puppeteers put on their own buggy show. About 20 different bugged-out exhibitors are on hand along with live bugs such as Giant African Millipedes, Madagascar Hissing Cockroaches and tropical walking sticks.

> **Contact:** 800-666-5397
> www.columbuszoo.org

Dragonfly Days

Valley Nature Center
Welasco, TX

Celebrate the world of "Mosquito Hawks" at this non-profit environmental education and conservation center based in a five-acre urban park. Discover the variety of odonates in the **Lower Rio Grande Valley** and learn how to distinguish between a Caribbean Yellow Face and Needham's Skimmer. Butterfly seminars, insect collecting field trips, and a banquet are included. A special children's program is held on Saturday.

Contact: 956-969-2475
www.valleynaturecenter.org
info@valleynaturecenter.org

> There are about 5,000 dragonflies and damselflies worldwide.

Plant Discovery Day

Ohio Agricultural Research and Development Center
Wooster, Ohio

Looking for rare and unusual plants? You'll find them at **Ohio State University's** annual event at its Wooster campus. Along with the flora comes fauna in the form of bugs at the **Biolab** for children featuring a bug zoo, a worm tunnel and leaf rubbings. The day includes wagon tours of the campus and **Secrest Arboretum**.

Contact: 330-263-3700
www.oardc.ohio-state.edu

Insect Fair

Skyline-Sanborn Park
Los Gatos, CA

Get close to amazing live arthropods introduced to visitors by the **Youth Science Institute**. The weeklong display of live bugs from butterflies to bees to millipedes mesmerizes young visitors. Kids observe insects in all stages of metamorphosis, take part in hands-on activities such as touching live silk worms and piecing together buggy puzzles. If the bugs aren't enough to give you the creeps, the thought that the famous **San Andreas Earthquake Fault** runs straight through the park, might.

Contact: Youth Science Institute Nature Center at 408-867-6940
www.ysi-ca.org/Sanborn/SBHome.html

June

Botanica's Butterfly Festival

Botanica, The Wichita Gardens
Wichita, KS

Spin into the opening of **Botanica**'s annual summer **Butterfly House** with this weekend festival devoted to papillons, mariposas and schmetterlinks. The words all mean "butterflies" and there will be hundreds of native ones flying free in the house. The festival often features the The Bug Lady and The Balloon Man, husband and wife performance artists from New York. They present the *Love Bug's Hug*, an interactive play that showcases The Bug Lady's hand-woven insect costumes.

Contact: 316-264-0448
www.botanica.org

Family Fun Fest

University of Alaska Museum
Fairbanks, AK

Every June the **East Lawn** and the galleries of this museum are crawling with kids and a few bugs to boot. It's a big day for the museum, which houses natural and cultural history collections. Children scatter throughout the museum for items in a scavenger hunt. Outside, in a large circus tent, several hands-on activities correspond with what the kids are supposed to find. Among the activities are bug exploration and a chance to meet creepy crawlies up close. Other fun at the fest includes making a sun print, dissecting owl pellets, sewing a Tlingit button blanket, casting fossils and panning for gold. In a smaller tent local American Indian tribes perform traditional native dances and games.

    Contact: 907-474-7505
          www.uaf.edu/museum/

Mt. Magazine International Butterfly Festival

Mt. Magazine, AR

Come see the beauty of the Diana Fritillary, a butterfly rarely seen elsewhere but bursting atop Mt. Magazine, the home of a new state park. Considered a living butterfly Eden, Mt. Magazine is populated by at least 91 of the state's 127 butterfly species. Thousands of visitors flock to the mountain's plateau to witness the flight of these butterflies and participate one of the largest butterfly festivals in the U.S. Inspired by the research of the Diana Fritillary conducted by world-renowned lepidopterist Dr. Gary Noel Ross, the two-day festival is full of family fun. Watch the **Dance of the Butterflies Parade**, listen to Bluegrass butterfly music, join a butterfly hike. Field guides of Arkansas' butterflies are available at the fest so you can take them on your butterfly adventures.

    Contact: 501-963-2244
          www.butterflyfestival.com

> Fairyflies can fit through the eye of a needle.

July

Butterfly Day

Outdoor Campus, Sertoma Park
Sioux Falls, SD

Flutter by this nature museum for its annual butterfly celebration. Take guided tours of the 6,000-square-foot outdoor butterfly garden, watch a butterfly puppet show, learn how to create your own backyard butterfly haven or even how to start your own butterfly raising business. Discover how to tag Monarchs for research, find out about **Sertoma Park**'s new butterfly house, and peruse a beautiful butterfly specimen collection. Usually held the second weekend in July, **Butterfly Day** includes educational presentations and fun nature crafts for kids.

Contact: 605-362-2777
www.outdoorcampus.org

BugFest

Daniel Stowe Botanical Garden
Belmont, NC

The bugs are bouncing at this 110-acre greenspace along the banks of **Lake Wylie** just 30 minutes from Charlotte. July is **Butterfly Month** at the garden with a host of buggy activities, particularly during the Saturday **BugFest**. Go on a bug safari, visit the walk-in insect cage and create incredible insect crafts. Every child takes home a small prize. Be sure to keep an eye out for butterflies—more than 1,000 can be seen in the garden on a good summer day.

Contact: 704-825-4490
www.dsbg.org

Honey Harvest Festival

Creative Discovery Museum
Chattanooga, TN

Make a beeline to this honey of a festival held the last Saturday in July. Local and national beekeepers and apiarists will buzz by to talk about the sticky business of hives and honeycombs. Watch honey harvesting, taste varieties of the golden nectar and roll beeswax candles. Discover the medicinal uses of honey, bee pollen and propolis. Find out how you can harvest the power of these pollinators by becoming a certified master beekeeper. Kids are busy as bees building beehives, designing windsocks and playing bee games.

Contact: 423-756-2738
www.cdmfun.org

BugFest

Magee Marsh Wildlife Area
Oak Harbor, OH

Complete the activities at all five bug stations and you'll get a "Certificate of Bugology" for going buggy. Use a net to catch butterflies at the **Butterfly Identification Station**, search for marsh bugs at the **Aquatic Insects Station**, play **Buggy Bingo** (with prizes) and **Create a Bug**. You get to munch on buggy snacks like barbecue and cheddar cheese larva at the **Edible Bugs Station**. The Ohio Department of Natural Resources, Division of Wildlife, sponsors the event at the **Migratory Bird Center** and gives out free Ohio Bugs posters. Be sure to spend time on the walking trails and check out the view from the observation tower.

Contact: 419-898-0960

Butterfly Bonanza

Chippewa Nature Center
Midland, MI

Bring a butterfly net for a guided butterfly hike at this 1,000-acre nature center's annual tribute to Michigan's butterflies. If you don't have your own, the center has a few to loan so you can take them with you on the 14 miles of trails. Discover the top 10 butterfly-attracting plants to put in your garden. Watch videos on the state's butterfly species and butterfly gardening. Kids can see a butterfly puppet show, hear storytelling and create their own butterfly crafts.

Contact: 989-631-0830
www.chippewanaturecenter.com

Insectomania

Museum of Discovery
Little Rock, AR

Sip succulent honey slushes at the annual weekend bug bash. The museum makes a honey of an opportunity to showcase its Honey Bee observation hive and live **Insect Zoo**. The Central Arkansas Beekeepers Association buzzes in with beekeeping demos. Join the Master Gardeners to learn how to attract butterflies to your garden, go on a bug safari, and make bugfully creative crafts. Discover how to tag a Monarch and admire the insect collection of 4-H members. Kids will love crawling through the fabric tunnels of a mock ant colony and coloring the ants on the walls of the exhibit area.

Contact: 501-396-7050 or 800-880-6475
www.amod.org

> Tarantulas defend themselves with their hairs, not their fangs.

Great Texas Mosquito Festival

Clute Community Park
Clute, TX

Texans are big on BIG and they've got big skeeters buzzing around this two-decades-old fest including Willie-Man-Chew, the world's largest mosquito. The 25-foot Willie is an inflatable Texasquito dressed up in cowboy hat and boots. More than 25,000 swatters visit the three-day festival about 55 miles southeast of Houston near the Gulf Coast. Enter the **Mr. & Mrs. Mosquito Legs Contest**, the **Ms. Quito Beauty Pageant** and the **Little Skeeter Baby Crawling Contest**. Taste spicy samples at the bar-b-que, fajita and rib cook-offs and buzz over to the rides at the carnival.

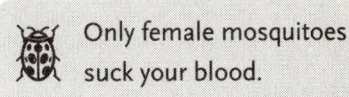

> Only female mosquitoes suck your blood.

Contact: 800-371-2971
   clutegtmf@computron.net

August

BuGFesT!

North Carolina Museum of Natural Sciences
Raleigh, NC

Ever been to the **Roachingham 500 Races**? Cheer on your favorite cucaracha in **Roachville** where you mingle with cockroaches of the world. The bugged-out community opens for a short time in a section of the museum's **Bicentennial Plaza** during the annual **BuGFesT!** Think you caught the biggest roach ever? Then enter it in the live **Triangle Cockroach Challenge**, which awards prizes for longest, heaviest, biggest and smallest roach. Only live ones are eligible.

Listen to live jazz from The Beetles, view ancient arthropods, observe a demonstration beehive, learn how to create a butterfly garden and get artsy with create-a-bug crafts. If you're hungry, try a few specialties from the **Café Insecta** like Real Men Eat Waxworms Quiche, Hoppin' Hunan Stir Fry with exotic Asian insects, Scuttling Scones and Ant Honeypot's Cheesecake. Try a bug full for free or munch on the traditional food and drinks for sale.

Contact: 877-4NATSCI or 914-733-7450
www.naturalsciences.org

Bug Fest

Southeastway Park
Indianapolis, IN

Earn your "Dr. of Bugology" degree in one afternoon when you attend Indiana's premier bug festival at this 188-acre park. There's so much to do, you may not have enough time to do it all. There are about 20 different insect stations with special activities. Kids can enter the **Bug Olympics'** six-legged race, play in the **Caterpillar Fun House** made of hay bug sculptures, and join an insect safari. Meet The Tarantula Lady, get the latest buzz from a beekeeper and be entertained by the **Bug Show**. Bob for aquatic bugs, walk through a butterfly tent and use sweep nets to scoop up insects in the field. Snack on chocolate fondue crickets, create ladybug fans and read buggy books at the bookmobile. The **Young Entomologists Society** and volunteers from the **Purdue University Entomology Department** fly in to answer bug-related questions.

Contact: 317-861-5167
www.indygov.org/indyparks/

Cleveland Metroparks Bug Fest

Garfield Park Nature Center
Garfield Heights, OH

Here's your chance to really go buggy—dress up as your favorite insect and win prizes in this fest's **Bug Costume Contest**. Thousands of visitors migrate here to see people cocooned in colorful insect garb and watch the fashion judges pick the best on Saturday afternoon. Earn a "Bugology Degree" at the two-day event by participating in several free activities. Race a mealworm, go on a bug hunt, play **Bug Bingo** or eat **Incredible Edible Insects**. The local **Mosquito Abatement** staff advises you on how to get rid of those bloodsucking pests or you can give blood willingly at the **Blood Mobile**. Talk to experts about other insects, too. Visit a live bee colony inside the nature center then look at the wildlife garden outside where they gather their pollen from flowering plants. The garden is also humming with birds and butterflies. Be sure to enter the buzzing black light forest created inside the center's library. Kids ages 6 to 14 can enter a buggy poetry contest a few weeks before the festival.

Contact: 216-341-3152
www.clemetparks.com/education/naturecenters.html

Festival of Butterflies

Powell Gardens
Kingsville, MO

Enjoy the bounty of beauty at the largest butterfly event in the world where 15,000 people gather to watch the butterflies blossom. The four-day festival at the 915-acre botanical bonanza flutters with activities for all ages. At the heart of the festival is the **Martha Jane Phillips Starr Butterfly Conservatory** where more than a 500 butterflies shimmer under the gardens' glass-domed conservatory. There's also the **Butterfly Breezeway**, a shade house featuring Monarchs and other locally raised butterflies. A giant topiary exhibit includes a 10-foot butterfly, a 10-foot caterpillar, a three-foot butterfly egg and a six-foot chrysalis. Kids activities include making a Monarch headband, participating in **Butterfly Catch & Release** sessions, a **Butterfly Expedition** (similar to a scavenger hunt) and a **Butterfly Coloring Contest**. World famous lepidopterists and horticulturalists present seminars, exhibits of exotic butterflies and photographs.

Contact: 816-697-2600
www.powellgardens.org

Hummingbird & Butterfly Day

Salman's Santa Fe Greenhouses
Santa Fe, NM

This nursery specializes in growing and selling butterfly and hummingbird-attracting plants and sponsors this event to help gardeners entice nature's winged jewels. Go on a free, guided onsite tour of **Xeriscape Demonstration Gardens** and take one of several seminars on butterflies and hummingbirds held throughout the day. Garden experts are available to answer questions and special plants are grown just for the festival. If the conditions are right, the festival will end with a stunning butterfly release.

Contact: 877-811-2700
www.santafegreenhouses.com

Bug Fest

Tualatin Hills Nature Park
Beaverton, OR

Bugs make the world go round at this 219-acre wildlife reserve's annual fest dedicated to arthropods. Learn about beetles, butterflies, bees, slugs and spiders and how nature's creepy crawlies help recycle fallen trees, pollinate flowers and provide food to larger animals. Play buggy games, touch live bugs, examine preserved specimens. Be sure to visit the interpretive center and look for bugs and butterflies along the five miles of trails.

Contact: 503-644-5595
www.thprd.org

Bug Day

Idaho Botanical Garden
Boise, ID

Get bugged at this 50-acre horticultural haven that was originally part of the **Idaho State Penitentiary. Albertson College** co-sponsors this adventure into the world of arthropods where you can earn your "Certificate of Bugology." Bring a bug for a professional entomologist to identify, play **Bug Bingo**, watch the **Insect Olympics**, catch live bugs and get a ladybug painted on your cheek. You can even snack on "edible insects." Be sure to visit the 13 specialty gardens, including a butterfly-hummingbird one.

Contact: 208-343-8649
www.idahobotanicalgarden.org

September

Bug Blast

Burke Museum of Natural History & Culture
Seattle, Washington

Located at the northwest corner of the **University of Washington** campus, this museum is definitely a blast to be at during their buggy event. Don't miss the museum's nine-foot-high **Bug Wall**, where you'll see 600 insect specimens entombed in a transparent casing. You can spy inside the honeycomb home of live bees, dig through compost for hidden bugs and inspect insect details under microscopes. There are tables and tables of live, fossilized and preserved bugs to see, touch or buy. Kids can get bugged-out by getting their faces painted, making bug catchers and maybe tasting a fried bug or two. If you've got a bug question or bugs you need identified, the members of the **Scarab Society** are there to help you.

Contact: 206-543-5590
www.washington.edu/burkemuseum

Insect-ival!

The State Botanical Garden of Georgia
Athens, GA

This 313-acre preserve at the **University of Georgia** boasts and roasts a variety of creepy and crawly critters at its bug festival. **Insect Discovery Stations** offer live insects for your inspection and cover topics such as Insect Friends and Foes, and Amazing Insects. Gourmet chef Lou Kudon demonstrates his culinary techniques using bugs as ingredients. Try a slice of Cricket Pizza or crunch Cricket Cookies for dessert. Other bugged-out cuisine is available at the **Insect Café**. Snack on tasty deep-fried mealworms after cheering on your favorite bug at the **American Roach Races** or the **Bess Beetle Races**. Children can earn an insect expert badge by participating in the **Insect-ival Discovery Hunt**.

Contact: 706-542-6156
www.uga.edu/botgarden/home2.html

Monarch Festival

Phipps Conservatory
Pittsburgh, PA

Say goodbye to summer with this Saturday festival celebrating the annual fall/winter migration of Monarch butterflies. Make Monarch headbands and egg-crate caterpillars inside the conservatory's **Gallery**. Watch a slide show of a Monarch's lifecycle and migration patterns in **Botany Hall**. Look for the **Touch the Caterpillar** discovery table where you can find out how to raise butterflies and learn how Monarchs are tagged for scientific tracking. Enjoy the last weekend of the year that **Discovery Garden**, an outdoor exploration garden with a giant tree stump and boxwood maze, is open. End the afternoon's festivities by watching a live release of butterflies.

>Contact: 412-622-6914
>www.phipps.conservatory.org

Monarch Butterfly Festival

Lake Katherine Nature Preserve
Palos Heights, IL

Celebrate the migration of the mighty Monarchs through the Midwestern prairies, at this 158-acre nature preserve along the **Calumet-Sag Channel**. Once a dismal dump, this wildlife oasis has a 20-acre lake, trails, waterfall, wetlands and wildflowers. The festival kicks off with a children's butterfly parade and has dozens of activities to keep everyone busy. Visit flying friends in the butterfly tent and some crawling ones in the spider tent. Taste a bug if you dare in the **Eat-A-Bug Challenge**. Listen to an insect expert give you the latest buzz about bugs then go for a creepy crawly hike in the **Buzz 'N Bloom Prairie**. Meet members of the **Young Entomologists Society (YES)** and watch aquatic bugs go swimming. Other activities include hayrides, canoe rides, children's crafts, scarecrow making, haystack treasure hunt, puppet shows, barn dancing and parachute games.

>Contact: 708-361-1873
>www.palosheights.org/lake/indexlk.html

The Great Insect Fair

Penn State
University Park, PA

Think you can go the distance? How about for spitting crickets? Then start juicing up those lips for the ultimate cricket-spitting competition. World champion cricket-spitter Dan Capps once set a record spit of

38 feet at this fair. Everybody can enter with senior and junior divisions for men and women. Kids can also test their might at the **Insect Olympics**. Under the **Bug Top** (a tent), kids compare their physical prowess to bugs by slipping into sacks and trying to crawl with their hands tied behind their backs like larvae would. Look for the Dancin' Bees (costumed adults) in a play at the **Lady Bug Theatre**, or pay a visit to the Bug Doctor. Get bugged out at the **Insect Zoo** featuring live Tobacco Hornworms, Australian Walking Sticks and Death's Head Cockroaches. Snack on insect hors d'oeuvres like Chocolate Chirpies (crickets) and Mealworm Stir-fry at the **Insect Deli**. Discover wondrous wings at the **Butterfly House** and learn how to plant your own butterfly garden. Meet insect collectors, start your own collection, watch a live beehive and sample fresh honey. Other kids' activities include story telling, bugly face painting and bugloons.

Contact: 814-865-1895
http://entscied.cas.psu.edu/insectfair.html

Bug Day!

Penitentiary Glen Reservation
Kirtland, OH

Imagine Alex Trebeck dressed up in antennae and you've got *Bug Jeopardy!* one of several bugged-out activities that visitors swarm to at this 386-acre park's day of bugs. Going on its 14th year in 2002, **Bug Day!** is always buzzing with something new. Everybody gets an activity card when they come to the event; get five ladybug-marked activities punched and you earn your "Bugology Degree." Play Bug Bingo, Bean Bag Bug Toss or crawl through an **Amazing Ant Maze**. Kids who love to create might not leave the craft tables where they can make butterfly kites, Doodlebug art or insect hummers. In a twist on musical chairs kids play the **Dung Beetle Ball**, work their way through metamorphosis in the **Worms to Wings Butterfly Obstacle Course** or challenge themselves on the team-building **Insect Adventure** course. Live critters are plentiful when the **Young Entomologists Society (YES)** brings its **Bugs On Wheels** featuring tarantulas, walking sticks, Praying Mantids and scorpions. Look for the giant wooden dancing disco ants and top-hatted, toe-tapping spider. They have cutouts for your family's smiling faces so you can snap photos of your own bugly brood.

>Contact: 440-256-1404
>www.lakemetroparks.com

Honey Fest

Indian Creek Nature Center
Cedar Rapids, IA

Iowa has a bounty of Honey Bees so a **Honey Fest** is a sweet way to honor these beneficial bugs. Each year the fest buzzes with activities and is so popular that even the Iowa Honey Queen pays a visit. Usually the 210-acre nature center is a peaceful place unless you walk along the **Bee Trail** where you hear the buzz of more than 25 active hives. A beekeeper takes you on a tour so you can see how nature's nectar becomes sugar for your tea. There are live demonstrations of beeswax candle-dipping and a beekeeper sports a beard of bees. Kids can roll their own candles and make crafts like a beaded bee key chain or pollinator magnets. Be sure to sip the honey-sweetened lemonade, snack on baked goods made with honey and peruse the varieties of honeys for sale.

>Contact: 319-362-0664
>www.cedarrapidsweb.com/places/cedarrapids/icnature/

Butterfly Festival

Arizona-Sonoran Desert Museum
Tuscon, AZ

Tour the beautiful bounty of butterfly gardens—in the desert—at this 100-acre wildlife wonderland. See how many of the 250 Sonoran

Desert butterfly species you can spot and get tips on how to observe the fluttering through binoculars. Docent-led butterfly walks are held every morning during the weeklong festival. The New Kiva Puppet Shows and crafts engage the kids. The museum shows videos and there are other live animals on display.

    Contact: 520-883-1380
        www.desertmuseum.org

Bug Bash

Folsom Children's Zoo and Botanical Gardens
Lincoln, NE

Ever see a roach pull a tractor? You'll have to see it to believe it at the festival that signals the end of the zoo's seasonal **Butterfly Pavilion**. Run for the **Roach Races** where American Cockroaches compete on AKSARBUG, Nebraska's premier **Cockroach Race Track**. In between you'll see the roaches pulling tractors. Meet with exotic tropical roaches while learning about roach biology and behavior. The roach stop is just one of more than a dozen bug stations that infests the zoo grounds. Tag a Monarch and release it to go on its migratory path, watch beekeeper trainees wrangle a hive of Honey Bees and sample their tasty nectar, visit Dr. Bug in his field hospital and help him diagnose patients exhibiting symptoms of bug bites. Fold colorful origami insects, listen to the music of Bo and the Weevils and see the Zoo Bug Players present a creepy-crawly drama. A special feature of the fest is collecting a set of **Bug Bash** trading cards. Show your master card at each station and you collect a new color card depicting one of Nebraska's insects with bug stats on the back.

> Honey Bees cool their hives by fanning water droplets.

    Contact: 402-475-6741
        www.lincolnzoo.org

Honey Bee Festival & Antique Show

Downtown Square & County Fairgrounds
Paris, IL

Thousands of visitors buzz into this weekend event that's busy with more than bees. Besides the **Bee Keeping Exhibit & Market** featuring an observation hive, honey and beeswax products, there's a carnival of family fun. Look for the **Civil War Encampment**, the antique tractor display and the **Honeybee Art Gallery** that showcases elementary student art. Watch a sheep rodeo, play Barnyard Bingo and participate in a cakewalk. Folk art demonstrations include spinning, weaving and whittling while children's activities include a petting zoo, walk through maze and barrel rides.

>Contact: Kiwanis Early Risers at 217-465-4176
>www.honeybeefestival.org

Insect Horror Film Festival

Iowa State University
Ames, IA

Like their buggy pals at the **University of Illinois**, entomology students at **Iowa State** host a fun fest featuring a bugged-out film. In 2001 visitors watched *James and the Giant Peach*. Before the film, go buggy with 90 minutes of bug-filled activities. Touch Madagascar Hissing Cockroaches and Giant African Millipedes at the **Insect Petting Zoo**. Enjoy the swirls of colors from shimmering wings inside a live butterfly tent and watch how bees build honeycomb in a live observation hive. What's a film without munchies? Sample the entomology department's specialties: Cricket Clusters and Mealworm Dip on Crackers. You can print out bugfully delicious recipes from the film fest's Web site.

>Contact: 515-294-7400
>www.ent.iastate.edu/entclub/horror/

Celebration Of Butterflies

Claiborne Parish Fairgrounds
Haynesville, LA

Visit the Butterfly Capital of Louisiana for a fluttering fun time. World-renowned lepidopterist Dr. Gary Noel Ross of Baton-Rouge jump-started this festival featuring a butterfly conservatory with live flutterbys. Take part in discussions with Dr. Ross, former director of butterfly festivals for the **North American Butterfly Association**. Go on wildflower walks, learn how to create the perfect butterfly garden and buy butterfly plants. Watch butterfly videos, view butterfly collections and create beautiful butterfly crafts.

Contact: Loice Kendrick at 318-624-1929
www.claiborneone.org/haynesville/

Insect Fair

Quail Botanical Gardens
Encinitas, CA

Held at the end of September, this fest is brimming with bugs from beetles to bees to ants to spiders. Exotic and native snakes are also on hand. Eat roasted bugs and snack on bug cookies, listen to bug story time and watch bugged-out videos. Go on a guided nature walk, get tips from bug experts and design crafty critters. There are lots of insect themed clothing and jewelry for sale even some bug-less treats!

Contact: 760-436-3036
www.qbgardens.com

Honey Bee Festival

Bishop Quinn High School
Palo Cedro, CA

For nearly 20 years this community near Redding has been buzzing with activity during its annual **Honey Bee Festival**. The event is so popular that even the American Honey Queen (crowned by the American Beekeeping Federation) pays a visit. Observe a real honey hive and learn how bees bee-have. Ride in a horse-drawn wagon, learn all about honey and have a honey of a time.

Contact: Palo Cedro Chamber of Commerce at 530-547-4554

West Virginia Honey Festival

Parkersburg City Park
Parkersburg, WV

Want to become a buzzy beekeeper? Then sign up for the **Bee School** at this community festival. Watch how honey is harvested and processed, sample sweet treats and buy the best bee products in West Virginia. Bee amazed as a beekeeper lets live bees form a beard on his face. Learn how to make beeswax candles and cook with honey. Enter the **Honey Baking Contest** and sample honeyed ice cream sundaes.

Contact: Great Parkersburg Convention and Visitor's Bureau
304-428-5835
www.parkersburgcvb.org

Monarch Butterfly Fiesta Day

Black Hill Regional Park
Boyds, MD

Celebrate the migration of the Monarchs at this vast regional park—it's more than 1,800 acres. Use nets to safely capture then release butterflies in the native wildflower meadow then take a walk through the butterfly garden to learn how to plant munchables for Monarchs. You get to take home a Monarch's favorite snack—a Milkweed plant—for your own garden. Explore a caterpillar rearing and butterfly free-flight enclosure, make Magical Monarch bathtub fizzies and meet The Magnificent Monarch Man. Sip Mexican hot chocolate and sample South of the Border treats while taking in the view of **Little Seneca Lake**.

Contact: 301-972-9396

October

The Incredible Edible Insect Event

Audubon Louisiana Nature Center
New Orleans, LA

Go bug-eyed over old-fashioned Southern-style cooking with a twist—everything is made with invertebrates, and we're *not* just talking crawdads or shrimp. This bug banquet features fun food such as Crispy Cajun Crickets, Sautéed Mealworms with mushrooms and Worm Fritters. Sample scrumptious Chocolate Chirp Cookies, Crawlines and Hush Buggies. Bring a toothpick, because those pesky fried crickets can get stuck between your teeth. Food samples are free. In 2002 a panel of celebrity judges crowned the best and most original dishes.

If you're not looking to challenge your taste buds, there's more to do. Touch a Madagascar Hissing Cockroach at the **Creepy Crawly Critter Petting Zoo** or become friends with Patent Leather Beetles, Vinegaroons, or millipedes. Go on a bug safari on the center's hiking trails where you look for indigenous insects. Enter the Roach Rally or Cricket Spitting Contest for extra buggy fun. Get a butterfly painted on your face, listen to bugged-out stories and create creature crafts. Be sure to take a stroll in the nature center's outdoor butterfly garden that attracts a variety of winged wonders including hummingbirds.

Contact: 800-774-7394
www.auduboninstitute.org/lnc
air@auduboninstitute.org

Edible Insects 2

Peabody Museum of Natural History
New Haven, CT

The worms are warm and the crickets are crisp at this annual feast for the fearless. Each year local chefs prepare a tasty hot insect fricassee and a special buggy dessert to delight the taste buds of those brave enough to try them. In 2001 even noted WNPR radio host Faith Middleton was inspired to fry up insects and serve them to the public.

The bug cook-off is not just to tantalize but also to educate visitors on the worldwide practice of entomophagy, also known as bug eating. Learn how insects can bring variety and taste to food, and about the historical, cultural and artistic aspects of insects as food.

Although insect-tasting is a highlight of the event, there's more bugfully delightful fun. Pet cool critters at the live insect zoo, get a bug painted on your face, make an insect craft and check out various insect displays. Bring in any weird bug, dead or alive, for **Stump the Entomologist** as an expert panel tries to identify your find. Visit the **Insect Art Gallery** featuring jewelry, drawings, paintings and sculptures with an insect theme.

Contact: 203-432-5050
www.peabody.yale.edu

Welcome Back Monarchs Day

Natural Bridges State Beach
Santa Cruz, CA

If it's October then the Monarchs have started flocking to the **Monarch Grove** at this beach park famous for its natural rock bridges. At least

100,000 migrating marvels spend the winter at their "city in the trees" in the Eucalyptus grove. Park docents conduct butterfly tours, offer educational Monarch programs and craft activities for kids. Meet super heroes Monarch Man and Monarch Woman, watch the migration parade, savor hand-cranked mariposa ice cream. Prizes are awarded to all attendees in costume.

Contact: 831-423-4609
www.cal-parks.ca.gov

Monarch Butterfly Migration Month

Rio Bravo Nature Center
Eagle Pass, TX

October is **Monarch Butterfly Migration Month** in this Texas town. The nature center conducts Monarch research and participates in Monarch tagging as a part of the international **Monarch Watch Program** and **Texas Monarch Watch**. Tags are provided free of charge to scouting troops and visitors can learn how to participate in the program. Help the nature center track the Monarchs migrating through the Middle Rio Grande region by reporting any "roosting" sites in your area.

Contact: 830-773-1836
www.riobravonaturecenter.org

Fire Ant Festival

Marshall, TX

Feeling antsy? Then this free fiery festival is for you. This tiny town gets so antsy with the red stinging insects that Oprah featured the fest on her TV show. Pick from a picnic of activities: **Fire Ant Calling Contest**, a **Fire Ant Round-up** where kids try to collect live ones in milk jugs, and "gurning," an ugly face contest where participants stick their heads through a Fire Ant decorated toilet seat and make the weirdest faces they can. Pedal in the **Tour de Fire Ant**, a bicycle tour (not race) that includes 10-mile, 35-mile, 45-mile and 100K courses. There's also a chili cook-off, but no one is willing to say if Fire Ants are part of the secret ingredients.

Contact: 800-953-7868
www.marshall-chamber.com

Honey Bee Festival
Hahira, GA

The American Honey Queen pays a visit to this annual fest that draws more than 29,000 visitors to this town near **Valdosta**. The event celebrates the Honey Bee, but has more than honey activities buzzing around town. There are carnival rides, carriage rides, 5-K runs, climbing walls, dancing and drill teams, gospel singers and quilting bees. The big finale is the afternoon parade through downtown.

Contact: 912-794-2567
www.hahira.ga.us/

Woollybear Festival
Victory Park
Vermillion, OH

Ohio's biggest one-day festival with attendance estimated at 100,000 celebrates the weather predicting abilities of the Woolly Bear. These fuzzy caterpillars turn into medium-sized Tiger Moths. Hosted by Cleveland's TV-8 weatherman Dick Goddard, the festival heralds the length and severity of the winter by examining the black caterpillars' orange stripes. The town puts on a big parade—the one in 2001 had a space theme complete with microgravity demonstrations. Cheer on your favorite woolly in the caterpillar races, participate in woolly games and dress up as a friendly Woolly Bear.

Contact: Sandusky/Erie County Convention & Visitors Bureau
800-255-3743
www.buckeyenorth.com

Cal Poly Pomona's Annual Pumpkin Festival

California State Polytechnic University
Pomona, CA

Pick a farm-raised pumpkin then take the tram from the pumpkin fields to the **Insect Fair**. The **College of Agriculture** sponsors the fair during the two-day **Pumpkin Fest**, featuring more than 40 education and commercial bug vendors. See live bees at the observation hive, buy fresh honey and honeycomb, sample insect treats such as chocolate-covered crickets and add beautiful beetles to your preserved bug collection. Look for bug books, buggy T-shirts and bugging supplies. The fair is popular so be sure to reserve enough time to get back on the tram; it gets crowded at the end of the day.

Contact: 909-869-3342
www.csupomona.edu

Creepy Crawly Fair

Descanso Gardens
La Cañada, CA

Every October the pumpkins start popping up at this 160-acre horticultural haven in the San Gabriel Valley and the bugs come creeping into the booths at **Van de Kamp Hall**. Examine a live observation Honey Bee hive, sample the sweet syrup and look over the live and preserved crawling critters. Pet a bug, watch wiggly worms and scuttling scorpions and create creepy crafts.

Contact: 818-952-4400
www.descansogardens.org

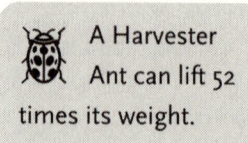

A Harvester Ant can lift 52 times its weight.

Banner Elk Woolly Worm Festival
Banner Elk, NC

Racing Woolly Worms, also known as Woolly Bears, is a serious business in **Banner Elk**, particularly since prize money is awarded to the fastest woolly. About 1,000 worm trainers prepare for the event that attracts more than 20,000 to this mountain town. The champion worm is chosen to be the official forecaster of the upcoming winter. Woolly Worms are really fuzzy caterpillars that turn into Tiger Moths and weather predicting lore says that you can forecast how harsh the winter will be based on the black larva's orange stripes. Check the Web site for the week-by-week predications by the winning caterpillar and the local weatherman.

Contact: Avery/Banner Elk Chamber of Commerce at 828-898-5605
www.woollyworm.com

The Ugly Bug Fair

Fullerton Arboretum
Fullerton, CA

Bugs are a big attraction at the annual **Arborfest**. About 6,000 people come to hear about worms and other wildlife, go on wagon rides, take historic house tours and pick pumpkins. Go on a bug safari with an arboretum nature guide, get up close to wriggly wonders from the **Insect Zoo** at the **Natural History Museum of Los Angeles County**, visit with **Dr. Sue's Traveling Insect & Arthropod Zoo** and meet a worm named Fred. There are also cactus and succulents for sale, scarecrow story telling, close encounters with real bats, flower potting, seed planting, butter churning and apple pressing.

Contact: 714-278-3579
www.arboretum.fullerton.edu

Texas Butterfly Festival
Mission, TX

Mission is a Mecca for butterflies and home of the biggest butterfly fest in the U.S. Located in the hub of the Lower Rio Grande Valley, the community is part of the most bio-diverse region in the country and home to about 300 butterfly species. Top lepidopterists and butterfly aficionados convene at this festival and offer seminars, hikes and field trips. Visit the temporary butterfly conservatory fluttering with live butterflies that you can hand feed. Line up in a butterfly or nature costume and join the band parading along the streets of Mission. Attend a butterfly book signing, a South Texas BBQ Dinner and find fun at the **Butterfly Bonanza**. A **Butterfly & Nature Expo** features educational and commercial vendors offering butterfly information and merchandise. Mission also has more than 480 species of birds so the festival attracts birders, too. Look for these butterflies seen at previous festivals: Guava Skippers, Mexican Bluewings, Crimson Patches, Violet Banded Skippers, Pixies and Silver Emperors.

Contact: Mission Chamber of Commerce at 800-580-2700
www.texasbutterfly.com/main.html
mission@missionchamber.com

Tarantula Festival

Coarsegold Historic Village
Coarsegold, CA

Trick-or-Treat with tarantulas at the annual fest held the last Saturday before Halloween. Treat yourself to a live tarantula derby, discover tips about tricky tarantula care from **Arachnida Club** members and compose a poem about your favorite eight-legged pal. There are several contests to enter: **Hairy Legs Contest, Scream Off, Children's Tarantula Race, Halloween Costume Contest Parade, Pumpkin Carving Contest, Best Pumpkin Dessert and Pizza Eating** (for kids ages 6 to 14). The day includes a cakewalk, gold panning, iron working, doll making, woodworking, beading and leather crafts. Arrange to take a guided tour of the town if you come with 10 or more people.

    Contact: Coarsegold Country Village at 559-683-3900
          www.coarsegoldca.com

November

Monarch Butterfly Release

Children's Discovery Center
San Jose, CA

The San Jose skies are all aflutter with orange and black colors when hundreds of Monarch butterflies are released at the largest children's museum in the West. The release is in partnership with **Magical Beginnings** of Los Gatos, which raises the Monarchs on a farm. The released butterflies are tagged and tracked for scientific research. Families can sponsor a butterfly for $15 if they purchase it in advance of the release date. Sponsorship also includes museum admission that day.

    Contact: (408) 298-5437
          www.cdm.org

South Carolina Insect Festival

Cypress Gardens
Moncks Corner, SC

Do the Grasshopper Hop at one of the biggest bug bops in the South. This all day fest brings out the best in bug activities. The Moncks Corner Dance Studio demonstrates the unusual grasshopper

> There are 10,000 different types of ants.

six-shoe so you can learn how to do it too. Then watch as the **Charleston Chinese School** performs traditional fan and butterfly dances. Take butterfly walks with a wildlife biologist, go on dragonfly hikes with an entomologist and go swamp dipping with a scientist. Kids will go buggy over the Clemson Entomology Club's bugfully delightful activities, the **Cockroach Races** and **Grasshopper Jumping Contests**. Brave bug lovers may want to sample insect cuisine or enter the **Metamorphosis Insect Costume Contest**. Get the latest bug buzz from the **Palmetto Insect Naturalists** and **South Carolina Beekeepers**. All free for kids 6 to 16 who come with an adult. Be sure to include a visit to the gardens' **Magic Wings Butterfly Conservatory** where 1,000 butterflies fly free.

Contact: 843-553-0515
www.cypressgardens.org

CHAPTER THREE

Bug Bytes

The buzz on the World Wide Web

● ● ● ● ● ●

The Web is a wonderful place to discover information *and disinformation*. There are thousands of buggy sites; I've plucked some of the best and present them here. They are, to my knowledge, reliable sources and as of this book's printing, the Web addresses are current. Look for updates on www.letsgobuggy.com. Happy Bug Surfing!

Ants

Myrmecology: The Scientific Study of Ants
http://www.myrmecology.org/

Everything you wanted to know about ants from anatomy to colonies to foraging. Perfect for school reports; includes an online bookstore. If you're really attracted to ants join the message board or chat with other ant aficionados.

Ants Online
www.antsonline.org

A highly interactive site about the structure of an ant. Choose your video resolution then click on it to enter the site. View it at low or high bandwidth or download it for offline viewing. Younger kids will need some help maneuvering the ant buttons.

Bees

University of Montana's Bee Alert
http://biology.dbs.umt.edu/bees/default.htm

How many bees does it take to equal the weight of one M & Ms™ candy? Find out the answer at this educational site that features live video of an observation hive. Every 15 seconds a camera records the bees' movements through their tube to the outside. Monitor forager flights and learn about the local Montana weather, too. The **Kids' Corner** has bee jokes, bee trivia, a coloring book of cartoon bees and information on beekeepers.

B-Eye
http://cvs.anu.edu.au/andy/beye/beyehome.html

See the world through a bee's eyes. Neuroscientist Andrew Giger set up this site so you can "see" how a bee would see other images. View how a bee sees another bee, a spider's web, even a picture of Einstein. Choose from a gallery of graphics, see what it's like to hover only 4 cm in front of an object or pick a spot on one of the patterns and zoom in.

Tales from the Hive
www.pbs.org/wgbh/nova/bees/

This *Nova* television site has excellent photos along with text explaining the life of bees. Discover how they make their honeycombs, eliminate predators like wasps, swarm around a queen and dance as a way of communication. There's a section on the different bee dances and you can watch them perform if you have QuickTime Player™. Learn how filmmakers managed to capture the incredible bee footage for the show.

The BeeHive: All About Honey Bees and Beekeeping
http://www.xensei.com/users/alwine/beesite.htm

Hives full of easy-to-navigate information—just click on the bees to learn about bees in winter, see photos of bees in a wall, learn how to start beekeeping. Offers some fun games: an interactive BeeHive crossword puzzle, Sting Me, Honey Bee jigsaw puzzle and Hang the Beekeeper (similar to the hangman word game).

Bug Cams

Ant Cam
http://www.antcam.com

Go straight to the source and watch streaming video of ants stacking eggs, moving dirt and foraging food. Two cameras are trained on the ant stations of Randy Rencsok, a computer scientist who has a keen interest in Myrmicinae, better known as ants. Randy built his own ant farms for the site.

Steve's Ant Farm
www.stevesantfarm.com

Steve runs his own Web site company, Web2Go, and this is a great example of what the company can do. Steve's got a minicam trained on an Uncle Milton's™ Ant Farm and you can see the ants digging their tunnels. View the ant farm during Eastern Time daylight hours; otherwise you get a black picture. The site includes an ant farm movie and scrapbook. You can even buy your own ant farm online.

The Butterfly Conservatory Butterfly Cam
www.amnh.org/exhibitions/butterflies/cams.html

During its annual winter butterfly exhibition, the **American Museum of Natural History** in New York City trains a live cam on a section of its butterfly conservatory. The picture instantly updates every few seconds depending on your computer's bandwidth. The site includes some pre-recorded QuickTime™ movies of butterfly feedings and emergence. The cam is live from 10 a.m. to 10 p.m.

Melbourne Zoo Butterfly Cam
www.zoo.org.au/static/Featured/webcams.cfm#

The clarity is wonderful on this live zoo cam. See a variety of colorful wings land on the feeding stations in the zoo's **Butterfly House**. Just remember that time in this part of Australia is 16 hours ahead of Eastern Standard time in the U.S. and 19 hours ahead of Pacific Standard Time. So when it's 6 p.m. on the East Coast and 3 p.m. on the West Coast it's 10 a.m. the *next day* in Australia.

Iowa State University Insect Zoo Cam
http://zoocam.ent.iastate.edu/

You have 120 seconds to participate on one of the Web's most active sites. In those two minutes you can operate a remote camera for a 360-degree view of whatever types of bugs are on exhibit. I've seen Assassin Bugs and Cave Roaches. Use the camera to zoom in on the best bugs.

University of South Carolina Roach Cam:
http://cricket.biol.sc.edu/usc-roach-cam.html

Check out these shots of Giant Madagascar Hissing Cockroaches. A new photo is transmitted every 60 seconds or so. Watch the mother roaches caring for their young. Look for a coin in the photo for size perspective; if you don't see it, the roaches have stolen it again! What's interesting, too, is the information on the older computer equipment used to bring these pictures to the Internet.

Draper's Super Bee Bee-Cam
www.draperbee.com/webcam/beecam.htm

Out in the Pennsylvania sticks is **Draper's Super Bee Apiaries, Inc.**, a company that provides bee supplies. They've got a camera on an observation hive 24 hours, 7 days a week. The picture updates every 30 seconds. If the picture is blank the company's dial-up connection is temporarily disconnected. However, you can view a short video clip buzzing with bee activity. Find out how to set up your own observation hive and get started in the bee business.

Bug Eating

Iowa State University's Tasty Insect Recipes
www.ipm.iastate.edu/misc/insectsasfood.html

If Jay Leno can eat Banana Worm Bread, Rootworm Beetle Dip and Chocolate Chirpie Chip Cookies, you might give these recipes a try. In addition to recipes for Chocolate-Covered Crickets, Mealworm Fried Rice and Corn Borer Cornbread Muffins, you'll get the lowdown on the nutritional aspects of bugs such as crickets, silk worms and Red Ants. Better yet, you can find out where to buy these insects to use as ingredients for your next gourmet meal.

Edible Insects
www.eatbug.com

Did you know that 100 grams of crickets has only 121 calories and 5.5 grams of fat? Compare that to ordinary hamburger with 288 calories and more than 21 grams of fat, and you might consider making a cricket croquet. Although this site on using bugs as food is no longer updated or maintained, it's still a good resource for an overview on bug eating, also known as entomophagy. Check out the great recipes for Mealworm Chocolate Chip Cookies, Chocolate Covered Crickets and Ant Brood Tacos.

Buggy E-mail Postcards (For Free!)

Milkweed Café
http://www.milkweedcafe.com/cards.html

Send virtual postcards with beautiful color butterfly photos. Just click on the photo, type in your message and send it off the e-mail recipient.

cards@ladybugcrossing.com
www.ladybugcrossing.com/cards/cards.html

Send a lucky ladybug to someone you love. There are five pages of colorful ladybug graphics to choose from. Add a holiday heading, music, a greeting and you'll quickly bring luck to an e-mail friend.

Insect Greeting Cards
www.bugpeople.org/cards/cards.htm

"Bug someone you love," with real insect photos. Send a close-up shot of bee or a bright butterfly.

Bug Fun For Kids

Katerpillars and Mystery Bugs
www.uky.edu/Ag/Entomology/ythfacts/entyouth.htm

From bug jokes to papier-mâché insects this University of Kentucky site is full of bug fun. Cook up bug food such as Ants-on-a-Log (celery, peanut butter and raisins) or learn how to use real insects as ingredients in food. Print out a buggy bookmark, create insect ornaments and design a compound eye. Good resource section for teachers and parents.

Ames True Temper Lawn and Garden Tools Kids Korner
www.ames.com/kids/index.html

Surf to this gardening tool supplier's site for the "I Know My Backyard" coloring book featuring 10 different insects, including five butterfly species, and several plants and flowers. The black-lined drawings are detailed yet easy to color when you print them out. The site also has a slider puzzle of a butterfly and flower; click on the pieces to move them.

National Honey Board
www.honey.com/kids

Get the sweet truth on honey and bees at the site that promotes the use of the syrupy nectar. A special kids' section buzzes with bee facts, honey history and lip-smacking recipes. Teachers can order **The Honey Files: A Bee's Life**, a combination 20-minute video and 96-page booklet for only $15.

Butterflies

The Butterfly Web Site
www.butterflywebsite.com

Visit the oldest butterfly Web site featuring Rick Mikula, butterfly expert, author and lecturer, who pioneered the **Butterfly Release Wedding**. The site is aflutter with a wealth of butterfly information including a **World Atlas** of butterflies, gardening tips, lists of public gardens, news and articles. Join a discussion group, browse a calendar of events and purchase butterfly related products from the online **Nature Store**.

Butterfly Zone
www.butterflies.com

If you want to bring bountiful beautiful butterflies into your backyard, this is the place to learn how. For those without a backyard, the site has a section for the urban gardener. The site's **Butterfly Guide** is divided into regional sections, so no matter where you live in the United States, you can find plants and gardening tips for attracting butterflies. **Butterfly Advisor** is a monthly column on how to be a better butterfly gardener and you can shop online for butterfly supplies, seeds, books and gifts.

Butterflies of North America
www.npwrc.usgs.gov/resource/distr/lepid/bflyusa/bflyusa.htm

Discover what butterfly species are found in your state at this **Northern Prairie Wildlife Research Center** site. Under the auspices of the **U.S. Geological Survey**, the research center monitors North American butterflies. Look for the photo thumbnails of butterflies to identify species where you live. Includes butterfly checklists for each county in your state.

Children's Butterfly Site
www.mesc.usgs.gov/resources/education/butterfly/Butterfly.shtml

Got a butterfly question? Check for the answers on this site run by the **Midcontinent Ecological Science Center**, also part of the **U.S. Geological Survey**. If your question isn't listed you can e-mail a butterfly expert. View a gallery of beautiful butterfly photos, print out coloring pages of the Monarch lifecycle and link to dozens of other educational butterfly Web sites.

Educational

Enchanted Learning
www.enchantedlearning.com

This is an educator's dream: a Web site saturated with printed educational materials and craft ideas. There are labeled printouts of more than a dozen insects. Just put "insect printouts" in the search box. Try inputting "butterflies," "ladybug" and "spiders" for hundreds of printables and ideas. The site highlights **Enchanted Learning Software** and seeks donations to keep it online.

Using Live Insects in Elementary Classrooms
http://insected.arizona.edu/lessons.htm

From the **University of Arizona Center for Insect Science** comes a site with complete lesson plans for teachers who want to use live insects in the classroom. Each lesson includes teacher preparation, actual lesson plans and additions such as vocabulary and a bibliography. Teach hygiene by using flies, discuss self-protection with the aid of grasshoppers and learn about nutrition by raising caterpillars into butterflies or moths.

The Activity Idea Place
www.123child.com/animals/bugs.html

Super easy instructions for bugs, bees and butterfly crafts from Sun Catcher Butterflies to Busy Bugs Headbands to Paper Plate Ladybugs. Directions for simple science and math games including Bee Stripe Math, Butterfly Wing Match and the Baby Bumble Bee song.

Insects

Insecta Inspecta World
www.insecta-inspecta.com

"The world is covered in bugs, so shouldn't you know a little bit about them?" This nice site is the creation of a group of students from **The Honors Academy of Thornton Junior High School** in Fremont, CA. Don't let the age of these Websters fool you. The site contains well-researched and easy-to-read articles on ants, bees, Scarab Beetles, Monarch Butterflies, crickets, fleas, Praying Mantids, malaria carrying mosquitoes and termites. Other interesting articles focus on bugs in art, bugs in the news and arachnophobia.

Alien Empire
www.pbs.org/wnet/nature/alienempire

Nova's flashiest bug site is a multimedia companion to the three-week *Nature* mini-series about insects shown on PBS. The site covers silk worms, Monarchs, mayflies, bees, Caterpillar Wasps and termites. There are videos you can download and an insect scramble to play. You need the Shockwave/Flash™ plug-in, RealPlayer™ and QuickTime programs.

Insectlopedia
www.insectclopedia.com

Essentially an online encyclopedia of insects. Look up species from A to Z, learn about medical research using insects, discover information on insect control, find schools that offer entomology and link to bug identification sites.

Monarchs

Monarch Lab
http://www.monarchlab.umn.edu

Help your class understand the basics of Monarch migration and biology at this site developed by **University of Minnesota** scientists and educators. Learn how to participate in the **Monarch Larva Monitoring Project** and how to raise Monarchs in the classroom. Site includes common questions and answers plus links to other Monarch, butterfly and education Web sites.

Journey North: A Global Study of Wildlife Migration
www.learner.org/jnorth

Students can register to participate in this free online study of seasonal change or anyone can use the site as an aid to understanding how and why certain wildlife migrates. Monarch Butterflies are one of the animal populations tracked as they switch locations based on seasons. Other animals include robins, hummingbirds, Bald Eagles, Gray Whales and manatees.

Roaches

Yucky Roach World
http://yucky.kids.discovery.com

It's a yucky world out there but somebody's got to tell us the facts about roaches. The **Yuckiest Site on the Internet** is the place to go. It's actually a clean fun site (think Nickelodeon™) with cartoon characters like Ralph Roach. Choose to enter with or without the Flash Plug-In to meet Ralph, read some amazing roach facts and get tips on roach removal.

Spiders

Arachnology For Kids
www.ufsia.ac.be/Arachnology/Pages/Kids.html

Too spooked by spiders? Then get caught in a Web of fun at this from the **University of Antwerp** in the Netherlands. There are about 100 links to spider sites for kids or for educators teaching students about arachnids. Learn how spiders spin silk, find out how some spiders are born swimmers and where to buy yummy chocolate spiders. Lots of links to elementary classroom projects, university sites and spiderific stories and books.

Tarantulas

Tarantulas.com
www.tarantulas.com

They're big, they're hairy and someone loves them on this site devoted to Theraphosidae and invertebrates. Check out the tarantula of the month, find out the answers to the most common tarantula questions and join the club (it's free). Buy tarantulas and tarantula food and get tips on tarantula care.

Tarantula Web Cam
www.museum.vic.gov.au/spidersparlour/tarant.htm

Welcome to **Spider's Parlour** at the **Museum Victoria** in Melbourne, Australia. A Spider Cam offers a continuous video feed of one of the tarantulas quarantined and off public view. The tarantulas were illegally imported by spider smugglers and found a home at the museum. Weekly tarantula feedings Friday at 3 p.m. Australian Eastern Standard Time (GMT +10 hours, GMT +11 hours summertime). Visit the rest of the parlour for information about the tarantulas and other arachnids.

Worms

The Adventures of Herman
www.urbanext.uiuc.edu/worms

Meet Squirmin' Herman the Worm. A cartoon character, he's the perfect spokesperson for young kids to learn about earthworms. Learn about his history, his family tree and anatomy. Click on the **Fun Place** for a coloring page, worm jokes and art gallery. Send in your own artwork to appear on the Web. Help Herman get to the soil's surface by playing the **Worm Tunnel** game. When you win you'll also have earned your "Wormologist Certificate," which you can print out.

Worm Composting Resources
www.wormwoman.com

Meet the wormiest woman on the Web, Mary Applehof. "Worms eat my garbage," says this composting queen who also publishes a book with that title. Mary champions the composting qualities of earthworms and provides resources to teachers and others who want to use worm power.

Worm World

www.yucky.kids.discovery.com

Just like Roach World, somebody's got to give us the dirt on earthworms and Wendell Worm, ace reporter, is just the bug to host that segment. Learn all about earthworms, visit Wendell's cousins and discover how worms are the Earth's recyclers. There's even a link to Worm Woman.

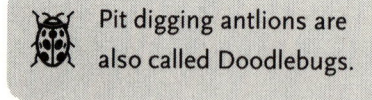

> Pit digging antlions are also called Doodlebugs.

CHAPTER FOUR

Be a Bugatist*

A host of buggy resources

● ● ● ● ● ●

Want to learn more about bugs? Do you know a child who wants to become an entomologist, an arachnologist or a lepidopterist? Are you an educator wanting to explore entomology? This chapter provides resources for really going buggy.

Here's a selection of buggy organizations, entomological suppliers, butterfly kits and suggested books to read.

*My daughter, Emelia Cassel, coined the word, meaning a person who studies bugs.

Bug & Butterfly Kits

Insect Lore
Box 1535, Shafter, CA 93263
800-LIVE-BUG, fax: 661-746-0334
www.insectlore.com

"Nature comes alive" with this company's easy-to-use kits for raising and observing butterflies, ladybugs, silk worms, earthworms, ants and spiders. Kits are available to individuals (we've used the **Butterfly Bungalow** and **Ladybug Lodge** with great success) and to schools and other large organizations. Order live larva, insect curriculum books and software. School and institutional orders must be made by mail or fax. Order the catalog or browse the colorful Web site.

The Nature Store
215-918-0729
www.thenaturestore.com

This online-only store has butterfly-raising kits and other bug and butterfly products. Buy live ladybugs, seeds for growing Milkweed (the Monarch's favorite food), binoculars, books and T-shirts.

Butterfly Gardening Supplies

National Gardening With Kids Store
800-538-7476
http://store.kidsgardening.com/nationalgardening

This **National Gardening Association Store** offers acres of kids' gardening supplies: composting materials, earthworm habitats and a selection of butterfly kits, games and books. Sign up for a free newsletter and print catalog. Check out the **Youth Garden Grant**—your school, community or civic organization might be able to fund a butterfly garden project.

High Country Gardens
2902 Rufina St., Santa Fe, NM 87507
800-925-9387
www.highcountrygardens.com

This is the mail order catalog of **Santa Fe Greenhouses**, a retail nursery specializing in butterfly, hummingbird and Xeriscape (water-wise) plants. Visit the Web site or request the beautiful 80-page color catalog offering 64 butterfly-attracting plants. Although matched to growing conditions in the West, many of the plants can be used elsewhere in the U.S. Sign up for the free e-zine: *Xeriscape Gardening News* and read past issues on the Web.

Entomology Suppliers

BioQuip
17803 La Salle Ave., Gardena, CA 90248
310-324-0620
www.bioquip.com

Started 55 years ago, this company offers a comprehensive entomology supply catalog for the professional entomologist and hobbyist. Their 200+ page catalog includes insect collecting equipment, preservation tools, microscopes and display cases. Educational materials include posters, puzzles, puppets, games, software and 1,600 book titles (including this one!).

Carolina Biological Supply Co.
2700 York Rd., Burlington, NC 27215
800-334-5551
www.carolina.com

An international supplier of science and math education materials, this company offers online resources for teachers. When you browse their Internet catalog, look under "Living Organisms" then click on "Animals." Here you'll find links to arachnids, crustaceans, habitats (with butterfly, millipede and tarantula homes), insects and insect cages. To order online you must sign up for a password and get credit approval.

Organizations

American Tarantula Society
P.O. Box 756, Carlsbad, NM 88221-0756
505-885-8406
www.atshq.org

Membership: $20 annually.

The world's largest arachnid society isn't just for people who love these hairy eight-legged beasts. It's also for those who have an interest in other arachnids. Membership brings you *Forum*, a magazine published six times a year. Be sure to visit their Web site for a gallery of photos and to access an archive of arachnid articles. If you've found a tarantula or just bought one on impulse, click on "Did you find a tarantula?" on the home page for a concise article on tarantula care.

Entomological Society of America
9301 Annapolis Rd., Suite 300, Lanham, MD 20706
301-731-4535
www.entsoc.org

Youth Membership: $10 annually.

The world's largest organization for professional entomologists offers youth memberships to students in grades K-12. Kids get a junior entomologist packet with either *Peterson's First Guide to Insects* or *American Entomologist*, a quarterly magazine appropriate for advanced middle school and high school students. Tap into the **ESA** Web site to discover what bug scientists are doing. **ESA**'s annual fall meeting is usually accompanied by a youth **Insect Expo**.

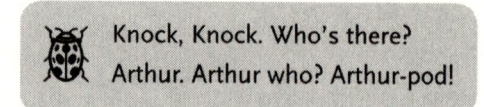

> Knock, Knock. Who's there?
> Arthur. Arthur who? Arthur-pod!

International Butterfly Breeders Association
P.O. Box 14102, Columbus, OH 43214
614-288-5677
www.butterflybreeders.com

Membership: $175 per company annually.

Interested in raising butterflies for sale or release? Then this is the organization to contact. The non-profit trade association has established guidelines for butterfly breeders to follow. Butterfly releases are under scrutiny by other organizations and by the **United States Department of Agriculture**, which governs butterfly-raising permits. This organization serves as a support group and clearinghouse for individuals and institutions that rear butterflies. Their Web site offers tips on how to plan the perfect butterfly wedding, information on where to buy butterflies and a livestock exchange.

Monarch Watch
University of Kansas, Entomology Program, 1200 Sunnyside Ave., Lawrence, KS 66045
888-TAGGING or 785-864-4441
www.monarchwatch.org

Membership: $25 annually, includes tagging kit.

The ultimate place to learn about Monarchs and their migration is this program and its outstanding Web site. Started by KSU professor Orley R. "Chip" Taylor a decade ago, non-profit **Monarch Watch** provides science education outreach, promotes Monarch conservation and involves thousands of students and adults in tracking the annual fall Monarch migration. **Monarch Watch** introduced butterfly tagging to trace their movements from Canada to Mexico. Learn how to join the tagging effort by visiting the Web. You can order additional tagging kits, Monarch butterfly-rearing kits, butterfly gardening kits and books about Monarchs.

National 4-H
www.4-h.org

Membership: No state or national dues

The **U.S. Department of Agriculture**'s county **Cooperative Extension**

Service operates a terrific youth program where kids ages 5 to 21 can learn about leadership, technology and livestock—even bugs! Dr. Art Evans, entomologist and research associate at the Smithsonian was a 4-H-er and studied insects in the program. To join, check the information on the Web site or look in the White Pages phone directory under county government for the nearest Extension office.

North American Butterfly Association
4 Delaware Road, Morristown, NJ 07960
www.naba.org

Membership: $30 regular, $40 family.

This non-profit organization works to promote the enjoyment and conservation of butterflies. Members receive *American Butterflies*, a stunning color quarterly magazine, and *Butterfly Gardener*, a quarterly publication, which includes a kids' page of activities. There are local chapters throughout the U.S. Check the Web site to find out if your state has one. **NABA** holds a **July 4th Butterfly Count** in an effort to create a yearly census of butterfly species. Volunteers pick an established count area with a 15-mile diameter and count the butterflies within that area for one day. The counts are held a few weeks before or after July 4th (July 1 Canada, September 16 Mexico).

Young Entomologists Society (YES)
6907 West Grand River Ave., Lansing MI 48906
517-886-0630
http://members.aol.com/YESbugs/bugclub.html

Membership: $12 Youth, $15 Adult, $18 Family, $20 Educator, $35 Junior Bugologist Kit, $35 Bugologist Kit, $50 Educator's Kit

One million served—that's the number of young minibeast enthusiasts that have been reached by this non-profit educational organization. Gary A. and Dianna K. Dunn run **Y.E.S.**, which now has its own **Minibeast Zooseum** in Lansing, MI (see Go Buggy chapter). Members get the *Y.E.S. NewsBulletin*, reduced rates on other **Y.E.S.** periodicals (there are several) and access to the Web site's members-only section. The membership kits come with goodies like certificates, pencils, bookmarks, activity books and a one-year-subscription to appropriate age level **Y.E.S.** periodicals.

Recommended Reading

The Family Butterfly Book
by Rick Mikula, ($16.95, Storey Books 2000). A great book for kids, families and schools to learn how to raise, care for and study butterflies. Includes a color identification guide for the 40 most common North American butterflies.

Pet Bugs, A Kid's Guide to Catching & Keeping Touchable Insects and
More Pet Bugs, A Kid's Guide to Catching and Keeping Insects and Other Small Creatures
by Sally Kneidel, ($12.95 each, John Wiley & Sons 1994, 1999). Perfect books for beginning bugatists. Tips on keeping crickets, snails, butterflies and more.

Monarch Magic! Butterfly Activities & Nature Discoveries
by Lynn Rosenblatt (Photographer) ($12.95, Williamson Kids Good Times!™ 1998). Filled with stunning color photographs. Forty Monarch activities, including how to raise, release and track the butterflies, keep kids busy.

Peterson First Guide to Insects of North America
($5.95, Houghton Mifflin 1998). For the junior bugatist, a concise guide to 200 of the most common insects in North America. Learn how to identify these insects and how to identify insect anatomy.

Totally Bugs
by **Pacific Science Center's** Dennis Schatz ($15.95, Silver Dolphin, 2000), is one of my favorite interactive bug books. Comes with a brightly illustrated 32-page book and 40 colorful plastic bug parts that kids can use to create 5 different bugs or mix them up to design their own creepy crawlies.

Play and Find Out About Bugs: Easy Experiments for Young Children
by Janice VanCleave ($12.95, John Wiley & Sons 1999). Learn how to

turn digging for worms and scouting for ladybugs into simple science with this award-winning science teacher and author. Team up with your child to conduct one or more of the 50 experiments with step-by-step instructions and easy-to-find materials.

The Eat-A-Bug Cookbook

by David George Gordon ($16.95, Ten Speed Press 1998). I've sampled naturalist David George Gordon's stir-fried crickets and orzo, and frankly, it was pretty good. (The crickets got stuck in my teeth though). This entomological epicurean guide is for the cook who can't wait to sauté a scorpion and for the diner who never imagined eating real mealworms.

Traveling Insects

Insect Safari

www.insectsafari.com

What has 18 wheels and flies? This free, multi-media expedition is rolling across the U.S. to bring insect education to schools and the public. The hands-on exhibit from the **Otto Orkin Insect Zoo** at the **Smithsonian National Natural History Museum** features 3-D models, classroom materials and interactive presentations that combine science, art and storytelling. Kids have to walk through a giant caterpillar to enter the **Insect Safari** mobile where they'll find insect specimens, diagrams and displays. The touring truck is set to visit 53 cities in 2002 alone, so sign up now to see if it can head to your school. **Insect Safari** is also sponsored by **Orkin Exterminating Company**, which funded the renovation of **Smithsonian's** insect zoo.

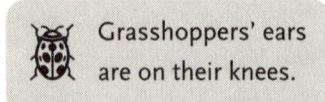

> Grasshoppers' ears are on their knees.

CHAPTER FIVE

Bug Bites

A guide to state insects and pet bugs

● ● ● ● ● ●

This chapter is for a bit of bug fun. Find out if your state has an official insect or butterfly or both! Discover the easiest bugs to keep as pets in your own insect zoo.

State Insects

All of the 50 U.S. states have official symbols such as flags, songs and seals. Some states have official birds, reptiles and mammals and 40 states have adopted an insect or butterfly as an official symbol. Of these, nine states have adopted an official state butterfly instead of *or* in addition to, an official state insect.

The Honey Bee is the official state bug in 15 states. In Tennessee, the Honey Bee is the official *agricultural* insect. The "Volunteer State" has two other state insects: the ladybug and firefly while the Zebra Swallowtail is designated as the state butterfly.

Does your state have an official insect or butterfly? If not, consider starting a campaign and lobbying your state legislature for an official insect or butterfly. Many of the state bugs got their status because elementary school students started petitions. The **Young Entomologists Society** has been active in trying to get the state of Michigan to adopt the Green Darner Dragonfly as a state symbol. To learn more about their campaign visit the **YES** Web site: http://members.aol.com/YESnetwk/.

These are all the state insects and butterflies as of this book's publication date. While some lists posted on the Internet differ from this, the *Let's Go Buggy!* list has been verified with each state government and or state library.

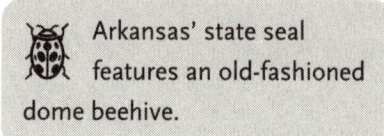

> Arkansas' state seal features an old-fashioned dome beehive.

Alabama: Monarch Butterfly (1989)
Alabama state *butterfly*: Eastern Tiger Swallowtail (1989)
Alaska: Four-spotted Skimmer Dragonfly (1995)
Arkansas: Honey Bee (1973)
California: California Dogface Butterfly (1929)*
Colorado: Colorado Hairstreak Butterfly (1996)
Connecticut: European Mantis "Praying Mantis" (1977)
Delaware: Convergent Ladybird Beetle "Ladybug" (1974)
Florida state *butterfly*: Zebra Longwing (1996)

Georgia: Honey Bee (1975)
Georgia state *butterfly*: Tiger Swallowtail (1988)
Idaho: Monarch Butterfly (1992)
Illinois: Monarch Butterfly (1975)
Kansas: Honey Bee (1976)
Kentucky state *butterfly*: Viceroy Butterfly (1990)
Louisiana: Honey Bee (1977)
Maine: Honey Bee (1975)
Maryland: Baltimore Checkerspot Butterfly (1973)
Massachusetts: Ladybug (1974)
Minnesota state *butterfly*: Monarch (2000)
Mississippi: Honey Bee (1980)
Mississippi state *butterfly*: Spicebush Swallowtail (1991)
Missouri: Honey Bee (1985)
Nebraska: Honey Bee (1975)
New Hampshire: Ladybug (1977)
New Hampshire state *butterfly*: Karner Blue (1992)
New Jersey: Honey Bee (1974)
New Mexico: Tarantula Hawk Wasp (1989)
New York: Nine-spotted Ladybird Beetle (1989)
North Carolina: Honey Bee (1973)
Ohio: Ladybug (1975)
Oklahoma: Honey Bee (1992
Oklahoma state *butterfly*: Black Swallowtail (1996)
Oregon: Oregon Swallowtail Butterfly (1979)
Pennsylvania: Firefly (1974)
South Carolina: Carolina Mantid (1988)
South Carolina state *butterfly*: Eastern Tiger Swallowtail (1994)
South Dakota: Honey Bee (1978)
Tennessee: Ladybug and Firefly (1975)
Tennessee state *agricultural insect*: Honey Bee (1990)
Tennessee state *butterfly*: Zebra Swallowtail (1994)
Texas: Monarch Butterfly (1995)
Utah: Honey Bee (1983)

Vermont: Honey Bee (1977)
Vermont state *butterfly*: Monarch (1987)
Virginia: Tiger Swallowtail Butterfly (1991)
Washington: Common Green Darner Dragonfly (1997)
West Virginia: Monarch (1995)
Wisconsin: Honey Bee (1977)

States without an insect or butterfly to call their own: Arizona, Hawaii, Indiana, Iowa, Michigan, Montana, Nevada, North Dakota, Rhode Island and Wyoming.

Basic Pet Bugs

Pill Bug Farm

Based on experience, Pill Bugs and Sow Bugs, sometimes called "Roly Polys" are just about the easiest bugs to keep around. These isopods don't require much: just moist soil stocked with dead leaves. Now how hard is that? We used a large plastic peanut butter jar to house our Pill Bugs. Be sure to puncture small holes in the lid for air. (We used a very thick nail and a hammer). Our Pill Bug colony has been reproducing for three years now.

Find out more at this **Oregon Zoo** Web site: http://zooscope.oregonzoo.org/TeacherSite/t12_pill_bugs.html and at this **University of Arizona Center for Insect Science** site: http://insected.arizona.edu/isoinfo.htm

Mealworms

Our second most successful bug colony has been mealworms which you can find at a fish bait store. We bought our first batch at an insect fair and raised successions of Darkling Beetles for almost two years. We used a simple plastic critter container from a pet store, filled it with oatmeal and kept slices of moist fruits and veggies inside. Apple worked best. Our only problem with our bug brood came when we experimented by using cornmeal. By doing so we invited moth larva to dine on the cornmeal and the moths have been difficult to get rid of ever since.

CHAPTER SIX

Bug Buzzwords

A glossary of bug biology

• • • • • •

Confused about the difference between an insect and a true bug? Want to know how to tell a butterfly from a moth? This simple glossary will help you dissect the bug buzzwords found in this book.

Apiarist: A person who keeps bees (beekeeper).
Arachnid: Has exoskeleton, 2 body parts, 8 legs, simple eyes—spiders, scorpions.
Arachnologist: Person who studies arachnids.
Arthropod: Segmented body, exoskeleton, jointed legs—insects, spiders, crustaceans.
Bug: A word kids use to describe any arthropod.
Butterfly: Insect with four colorful wings, antennae that look like miniature Q-Tips,™ wings are erect when at rest, usually flies during the day.
Caterpillar: Insect larva usually associated with butterflies and moths.
Chrysalis: Casing formed when caterpillars (larva) turn into pupa before emerging as butterflies (adults).
Cocoon: Casing formed when caterpillars (larva) turn into pupa before emerging as moths (adults).
Entomology: The scientific study of insects.
Entomologist: A scientist who studies insects.
Entomophagy: The eating of bugs.
Eusocial: Insects living with 2 or more generations divided into workers and reproducers—all ants, termites and some bees and wasps.
Exoskeleton: Skeleton located on body's outside—insects, arachnids, crustaceans.
Incomplete Metamorphosis: Three life stages: egg, nymph and adult. Babies look like miniature adults.
Insect: Body divided into three parts—head, thorax, abdomen. Has an exoskeleton. Six legs attached to thorax.
Invertebrate: Animal without a spine or backbone.
Larva, larvae (pl.): Metamorphic stage between egg and pupa.
Lepidoptera: Butterflies and moths.
Lepidopterist: A person who studies butterflies and moths.
Metamorphosis: Four life stages: egg, larva, pupa and adult.
Moth: Insect with four wings, less colorful than butterflies, feathery antennae, wings are flat when at rest, usually flies at night.
Pupa, pupae (pl.): Metamorphic stage between larva and adult. Chrysalis for butterflies, cocoons for moths.
True Bug: Insect with front wings that look like "half wings."

Index

A Bug's Life, 21
Abilene Zoo and Discovery Center, 159
Academy of Natural Sciences, The, 140
Adirondack Park, 118
Albertson College, 205
Albuquerque Biological Park, 109, 110
Allyn Museum of Entomology, 65
American Museum of Natural History, 63, 111, 226
American Tarantula Society, 238
Arachnamania!, 78
Arboretum of Los Angeles County, 191
Arborfest, 219
Arcadia Insect Fair, 191
Arizona-Sonoran Desert Museum, 18, 209
Arthropod Museum, 25, 189
Arts & Science Center, 124
Ashland Nature Center's Butterfly House, 52
Audubon Insectarium, 79
Audubon Louisiana Nature Center, 79, 80, 214
Audubon Nature Institute, 79
Aventis CropScience Insectarium, 119
Backyard Butterfly Lab, 140
Backyard Wildlife Center, 155
Ballet of the Butterflies, The, 162
Ballou Park Nature Center, 165
Banner Elk Woolly Worm Festival, 219
Belle Isle Zoo, 90
Big Bugs, 70
BioBlitz, 45, 46
Biophilia Nature Center, 14
BioQuip, 192, 237
Bioscape, 167

BioWorks, 59, 60
Bishop Quinn High School, 212
Blachly Conservatory, 162
Black Hill Regional Park, 213
Blooming Butterflies, 178
Blooms & Butterflies, 130
Bolz Conservatory, 178
Botanica, The Wichita Gardens, 77, 196
Botanica's Butterfly Festival, 196
Bristow Butterfly Garden, 169
Broadway at the Beach, 145
Bronx Zoo, 113
Brookfield Zoo, 71
Brookside Gardens, 81
Bug Bash!, 101
Bug Bash, 210
Bug Blast, 205
Bug Bowl, 187
Bug Day!, 208
Bug Day, 193, 194, 205
Bug Fest, 24, 202, 204
Bug House, 21
Bug House, The, 88, 91
Bug World, 172
Bug Zoo, 24, 108
BuGFest!, 123, 201
BugFest, 188, 194, 198, 199, 202
Bugs Alive, A Living Arthropod Exhibit, 139
Bugs and Other Insects, 116
Bugs!, 154
Bugz All Day, 189
Bumble Bee Discovery Kiosk, 101
Bumble Boosters, 101

249

Burke Museum of Natural History and Culture, 205
Butterflies & Blooms, 172
Butterflies & Orchids, 31, 185
Butterflies Forever, 138
Butterflies in Flight, 100
Butterflies of Alabama, 15
Butterflies of Alaska, 16
Butterflies of the Central Coast, 34,
Butterflies USA, 84
Butterflies!, 37, 71
Butterflies, 46, 140
Butterfly Adoption & Eco-Arts Festival, 188
Butterfly Bonanza, 200
Butterfly Conservatory, The, 111
Butterfly Day, 150, 198
Butterfly Festival, 209
Butterfly Forest, 143
Butterfly Garden, 136, 168
Butterfly Habitat Garden, 50
Butterfly House, 77, 88, 129, 132, 141, 147, 196
Butterfly Jungle, 73
Butterfly Kingdom, 148
Butterfly Landing, 84, 85
Butterfly Pavilion & Insect Center, 40
Butterfly Pavilion, 101, 121, 145, 210
Butterfly Place, The, 82, 100
Butterfly Sanctuary, The, 56
Butterfly Station, 165, 166
Butterfly World, 35, 53, 54
Butterfly Zone, 113
Butterfly/Hummingbird Garden, 89
Buttinger Nature Center, 107
Cal Poly Pomona's Annual Pumpkin Festival, 218
California Science Center, 28
California State Polytechnic University, 218
Callaway Gardens, 10, 66
Camden Children's Garden, 105
Carolina Biological Supply Co., 237
Carolina Pavilion, 120
Catawba Science Center, 124
Cecil B. Day Butterfly Center, 66

Celebration of Butterflies, 212
Center for Lepidoptera Research, 65
Central Park Zoo, 114
Charlotte Nature Museum, 121, 122
Chicago Academy of Sciences, 68
Chicago Botanic Garden, 70
Children's Discovery Center, 221
Children's Garden, 70
Children's Museum of Maine, 80
Chippewa Nature Center, 200
Cincinnati Zoo & Botanical Garden, 127
Claiborne Parish Fairgrounds, 212
Cleveland Botanical Garden, 134
Cleveland Metroparks Bug Fest, 202
Cleveland Metroparks Zoo, 132, 133
Cockrell Butterfly Center, 157
Columbus Zoo, 194
Community Days, 20
Connecticut State Museum of Natural History, 45
Cox Arboretum, 132
Crawl Space, The, 128
Crawl-A-See-'Em, 41
Crawl-a-seum, 151
Creative Discovery Museum, 10, 152, 199
Creepy Crawly Fair, 218
Critters and Things, 62
Cypress Gardens, 55, 147, 221
Dakota Zoo, 124
Dallas Museum of Natural History, 162, 186
Daniel Boone Butterfly Palace, The, 39
Daniel Stowe Botanical Garden, 198
Danville Science Center, 165
Day Butterfly Center, 66
Delaware Nature Society, 52
Denver Zoo, 42
Descanso Gardens, 218
Detroit Zoo, 89
Discovery Center, 124, 125, 180
Discovery Place, 121, 122
Discovery Room, 102
Dragonfly Days, 195
Draper's Super Bee Apiaries, 227

Edible Insects, 2, 46, 47, 215
Eleanor Armstrong Smith Glasshouse, 134
Entomological Society of America, 238
Epcot Center, 190
Exploration Station, 29
Explore Floor, 80
Exposition Park, 25, 27, 28
Fairbanks Museum & Planetarium, 164
Family Festival Bug Day, 186
Family Fun Fest, 196
Faust Park, 98
Festival of Butterflies, 203
Field Museum, The, 72
Fire Ant Festival, 216
Florida Museum of Natural History, 65
Flying Rainbows Café, 84
Folsom Children's Zoo and Botanical Gardens, 101, 210
Foremost's Butterflies Are Blooming, 86
Fort Wayne Children's Zoo, 73
4-H, 200
Frances R. Redmond Conservatory, 83
Frances V.R. Seebe Tropical America, 110
Franklin Park Conservatory & Botanical Garden, 130
Franklin Park Zoo, 84
Frederik Meijer Gardens, 86
Fullerton Arboretum, 219
Garfield Park Nature Center, 202
Good Zoo & Benedum Planetarium, 175
Grace Jarecki Seasonal Display Greenhouse, 87
Grand Rapids Children's Museum, 87
Great Insect Fair, The, 207
Great Texas Mosquito Festival, 201
Greathouse Butterfly Farm, 63, 64
Guinness Book of World Records, 96, 187
Gulfcoast Wonder and Imagination Zone (G.Wiz), 58
Habitat, The, 58
Hamill Family Play Zoo, 71
Harrell Discovery Center, 60
Henry Cowell State Park, 193
Henry Doorly Zoo, 102

Henry Vilas Zoo, 180
Heritage Park Zoo, 22
Heritage Pest Control, 23
Hershey Gardens, 141
Hidden Jungle, The, 30, 31, 32, 186
High Country Gardens, 237
Honey Acres, 176
Honey Bee Festival & Antique Show, 211
Honey Bee Festival, 212, 217
Honey Fest, 209
Honey Harvest Festival, 153, 199
Honey of a Museum, 176
Houston Museum of Natural History, 157, 184
Hug-a-Bug, 184
Hummingbird & Butterfly Day, 204
Huntsville-Madison County Botanical Garden, 16
Iceworm Festival, 185
Idaho Botanical Garden, 205
Idaho State Penitentiary, 205
Ijams Nature Center, 153
Incredible Edible Insect Event, 80, 214
Incredible Invertebrates, 15
Indian Creek Nature Center, 209
Indonesia Rain Forest, 73
Insect Fair, 192, 195, 212
Insect Fear Film Festival, 184
Insect Festival, 189
Insect Gallery, 181
Insect Horror Film Festival, 211
Insect Lore, 236
Insect Safari, 49, 242
Insect Zoo, 36, 74, 75, 137
Insectarium, The, 142
Insect-ival!, 206
Insectomania, 200
Insects: 105 Years of Collecting, 72
International Butterfly Breeders Association, 239
International Flower & Garden Festival, 190
Invertebrate Inn, 153
Invertebrates in Captivity Conference, 20

Iowa State University, 74, 211, 226, 227
Iron Butterfly Garden, 84
Joan Stifel Corson Butterfly and Wildflower Gardens, 174
John Hampson's Bug Art, 164
Judy Istock Butterfly Haven, 68
Kalamazoo Nature Center, 91
Kansas State University Butterfly Conservatory & Insect Zoo, 75
Kansas State University Garden, 75
Kate Gorrie Butterfly House, 107
Katydid Insect Museum, 22
KidSpace Museum, 188
Kirby Science Discovery Center, 151
Kirkwood Gardens, 104
Knight Rain Forest, 121
Knoxville Zoo, 154
Krohn Conservatory, 131
Lake Katherine Nature Preserve, 207
Las Palmas Park Butterfly Garden, 40
Las Vegas Natural History Museum, 103
Le Chateau des Papillons, 160
Lena Meijer Conservatory, 86
Lexington Children's Museum, 189
Lexington Community College, 189
Liberty Science Center, 108
Lichterman Nature Center, 155
Living Conservatory & Arthropod Zoo, 122, 123
Los Angeles Zoo, 28
Louisville Zoo, 10, 78
Lowry Park Zoo, 60
Mackinac Island Butterfly House, 93
Magee Marsh Wildlife Area, 199
Magic Wings Butterfly Conservatory, 83, 119, 222
Magical Beginnings, 221
Magical World of Butterflies, The, 131
Mariposa de Carmel, 34
Martha Jane Phillips Starr Butterfly Conservatory, 203
Mary Ann Lee Butterfly Wing, 98
Melbourne Zoo, 226

Miami Museum of Science, 57
Michigan Butterfly Garden, 87
Michigan State University, 88
Midcontinent Ecological Science Center, 230
Middletown Butterfly Farm, 144
Milwaukee Public Museum, 177
Minibeast Zooseum and Education Center, 92, 124
Minnesota Zoo, 95
Monarch Butterfly Festival, 207
Monarch Butterfly Fiesta Day, 213
Monarch Butterfly Migration Month, 216
Monarch Butterfly Release, 130, 221
Monarch Festival, 206
Monarch Program Butterfly Vivarium, 33
Monarch Sanctuary, 159
Monarch Watch, 77, 150, 162, 216, 239
Monsanto Insectarium, 96
Montana State University, 188
Montgomery Zoo, 15
Montshire Museum of Science, 163
Moody Gardens, 156, 157
Mosquito Hill Nature Center Butterfly House, 179
Mt. Magazine International Butterfly Festival, 197
Mung Juice, 166, 167
Museum Days, 139, 190
Museum of Discovery, 24, 200
Museum of Life and Science, 119
Museum of Science & Industry (MOSI), 59
Museum Victoria, 232
Museums on the River District, 77
National 4-H, 239
National Gardening Association, 236
National Mall, 194
National Museum of Natural History, 47, 194
National Pest Control Month, 94
National Zoo, 49
Native Species Butterfly House, 118
Natural Bridges State Beach, 215
Natural History Museum of Los Angeles County, 6, 8, 25, 27, 28, 192, 219

Natural Treasures: Past and Present, 32
Naturalist Center, 124
Nature Store, The, 229, 236
Nature Walk, The, 60
Nature Works, 151
Newport Butterfly Zoo, 144
Norfolk Botanical Gardens, 169, 170
North American Butterfly Association, 240
North Carolina Museum of Natural Sciences, 122, 201
Northern Prairie Wildlife Research Center, 229
Obee Rd. Garden Center, 129
Oglebay Family Resort, 175
Oglebay Institute, 174
Ohio Agricultural Research and Development Center, 195
Ohio State University, 194, 195
Oklahoma City Zoo, The, 136
Olbrich Botanical Gardens, 178
On Wings of Wonder, 67
Oregon State University, 139, 190
Oregon Zoo, 137, 246
Otto Orkin Insect Zoo, 6, 47, 48, 191, 194, 242
Outdoor Campus, 149, 150, 151, 198
Pacific Science Center, 6, 170, 241
Panhandle Butterfly House, 60, 64
Parkersburg City Park, 213
Pavilion of Wings, 26, 27
Peabody Museum of Natural History, 46, 215
Peggy Notebaert Nature Museum, 68
Penitentiary Glen Reservation, 208
Penn State, 207
Philadelphia Eagles Four Seasons Butterfly House, 105
Phipps Conservatory & Butterfly Gardens, 143, 206
Phoenix Zoo, 21
Pink Palace Family of Museums, 155
Plant Discovery Day, 195
PNM Butterfly House, 109
Pollinarium, 49
Powell Gardens, 203
Puelicher Butterfly Wing, 177

Purdue University, 187, 202
Quail Botanical Gardens, 212
Quinlan Visitor Center, 76
Rainforest Pyramid, 156
Rainforest, The, 133
Ralph J. Lamberti Tropical Forest, 115
Ralph M. Parsons Insect Zoo, 6, 25
Randall Museum, 193
Red River Zoo, 126
Rio Bravo Nature Center, 216
Rio Grande Botanic Garden, 109
Rio Grande Zoo, 110
Roaring Brook Nature Center, 45
Saint Louis Zoo, 10, 96
Salman's Santa Fe Greenhouses, 204, 237
San Bernardino County Museum, 6, 29
San Diego Natural History Museum, 32
San Diego Wild Animal Park, 30, 185
San Francisco Zoo, 36
Sarasota Jungle Gardens, 62
SASI, 20
Scarborough Faire the Renaissance Festival, 160, 161
Schrader Environmental Education Center, 174
Science Center of Connecticut, 44, 45
Science Museum of Minnesota, 94
Science Museum of Virginia, 166, 167
Sciencenter, 117
Sertoma Club Butterfly House, 150
Sertoma Park, 124, 149, 150, 198
Sioux Falls Butterfly Garden, 149
Six Flags Marine World, 35
Skyline-Sanborn Park, 195
Smithsonian National Museum of Natural History, 6, 47, 50, 194, 242
Smithsonian, 6, 47, 49, 191
SoCal Field Guides, 40
Sonoran Arthropod Studies Institute, 19
Sophia M. Sachs Butterfly House and Education Center, 98
South Carolina Insect Festival, 221
Southeastway Park, 202

Southwest Florida Water Management District, 59
Squam Lakes Natural Science Center, 104
St. Louis Zoo, 10, 96
State Botanical Garden of Georgia, The, 206
State Fair of Texas, 162
Staten Island Children's Museum, 116
Staten Island Zoo, 115
Steve's Bug Off Exterminating Company, 142
Stony Brook-Millstone Watershed Association, 107
Tarantula Festival, 11, 220
Tarantula Grotto, The, 22
Tessman Butterfly House, 16
Texas Butterfly Festival, 11, 220
Texas Discovery Gardens, 162
Texas Monarch Watch, 216
Texas Wild: Animals Alive!, 159
Toledo Zoo, 128
TOTE Family Fun Fest, 17
Tradewinds Park, 53
Tropical American Rain Forest, 136
Tropical Butterfly House & Insect Village, 170, 171
Tropical Discovery, 42
Tropical Forest, 115
Tualatin Hills Nature Park, 204
Tucson Mountain Park, 19
Tulsa Zoo, 135
Turtle Bay Museum, 37
Ugly Bug Fair, The, 219
Under the Bridge on the Astoria Riverfront, 138
Underground Adventure, 72
University of Alaska Museum, 17, 196
University of Antwerp, 232
University of Arizona Center for Insect Science, 230, 246
University of Arkansas, 25, 189
University of Connecticut, 45
University of Georgia, 206
University of Illinois, 184, 211
University of Kentucky, 189, 228
University of Minnesota, 231
University of Montana, 224
University of Nebraska, 101
University of South Carolina, 226
University of Washington, 205
University of Wyoming Insect Museum, 181
Upward Bound Butterfly Garden, 57
UTC Wildlife Sanctuary, 44
Valley Nature Center, 195
Virginia Living Museum, 168
Walt Disney World, 190
Washington Pavilion of Arts and Sciences, 151
Welcome Back Monarchs Day, 215
West Virginia Honey Festival, 213
Western Colorado Botanical Gardens & Butterfly House, 43
Western Colorado Botanical Society, 43
What's the Buzz?, 87
Wildlife Conservation Society, 114
Wildlife Interpretive Gallery, 89
Wings of Fancy, 81
Wings of Wonder, 135
Wings of Wonder, The Butterfly Conservatory, 55
Winnick Family Children's Zoo, 28
Witte Museum, 159
Woodland Park Zoo, 172
Woolly Worm Festival, 11, 219
Woollybear Festival, 217
World of Life, 28
World of Reptiles, 67, 68
World of Spiders, 90
World of the Insect, 127, 128
Yale University, 46
Young Entomologists Society, 92, 202, 207, 209, 240, 244
Youth Science Institute Nature Center, 196
Youth Science Institute, 195
Zoo Atlanta, 67
Zoolab & Butterfly Garden, 95, 96
Zooseum, 92

Want to keep going buggy?

Then order more copies of *Let's Go Buggy!*

It's easy. Just fill out the form and mail it in.

☐ **YES**, I want _____ copies of *Let's Go Buggy! The Ultimate Family Guide to Insect Zoos & Butterfly Houses*, at $14.95 each, plus $3.95 shipping per book. (California residents please add $1.23 sales tax per book). Allow 15 days for delivery.

My check or money order for _____ is enclosed.

Name _____

Organization _____

Address _____

City/State/ZIP _____

Phone _____ Fax _____

E-mail _____

Please make your check or money order payable to:

Corley Publications
P.O. Box 16969
Encino, CA 91416-6969

Fax: 1-877-376-2668
E-mail: booksales@letsgobuggy.com